INTERNATIONAL CENTRE FOR MECHANICAL SCIENCES

COURSES AND LECTURES - No. 130

OLGIERD C. ZIENKIEWICZ
UNIVERSITY OF WALES, SWANSEA

INTRODUCTORY LECTURES ON THE FINITE ELEMENT METHOD

COURSE HELD AT THE DEPARTMENT OF
MECHANICS OF SOLIDS
JULY 1972

UDINE 1972

Springer-Verlag Wien GmbH

Originally published by Springer Verlag Vien New York in 1972

ISBN 978-3-211-81202-0 ISBN 978-3-7091-2973-9 (eBook)
DOI 10.1007/978-3-7091-2973-9

PREFACE

In presenting the introductory lectures to an audience of engineers of a "mixed kind" including some for whom the subject is new and others already well versed in its application an obvious difficulty occurs. For the first, a complete coverage of essentials is necessary at the same time touching on a variety of problems of practical interest. For the latter, an expectation of some novelty beyond that available in texts and published papers is necessary to indicate some recent lines of thought. Further the whole presentation has to be compact and limited in space and time. It is within this framework that the present notes were prepared.

As many of the fundamentals and also some quite complex applications are presented in the author's text (which is referred to as reference A), a "telegraphic" style is adopted condensing some of the essentials to the types of notes that can be taken by a student in a lecture course.

Frequent reference is made to "reference A" and indeed further explanation, elaboration and proof is often presented there. While the larger part of the notes covers standard material, several points are made which give material either not available in the text or which is still subject of current research.

While "reference A" contains a wide bibliography, specific references are extracted at the end of these lectures either to draw attention to classic papers or for an elaboration of some recent researches.

Olgierd C. Zinkiewicz

Udine, July 1972

Historical Note-Discrete Analysis

Basis of discrete engineering analysis (e.g. structural mechanics) is:

1. Isolation of individual members (elements)
2. Establishment of two sets of quantities $\underset{\sim}{u}^e$ and $\underset{\sim}{F}^e$ associated with each element which have the property
   a) that if $\underset{\sim}{u}$ stands for global values

$$\underset{\sim}{u}_i = \underset{\sim}{u}_i^e$$

   b) that $\sum \underset{\sim}{F}_i^e = 0$, gives the second physical requirement
   c) that for each element we can determine the relation

$$\underset{\sim}{F}^e = \underset{\sim}{F}^e(\underset{\sim}{u}^e).$$

These three relations (together with appropriate physical constraints) allow a set of equations for $\underset{\sim}{u}$ to be found from which this can be determined.

In context of linear structures we have for example

$\underset{\sim}{u}^e$ displacement of nodes

$\underset{\sim}{F}^e$ forces exerted by each element

and (a) and (b) ensure interelement continuity and equilibrium respectively, (c) is a linear relation

$$\underset{\sim}{F}^{e} = \underset{\sim}{k}^{e} \underset{\sim}{u}^{e} + \underset{\sim}{F}_{0}^{e} .$$

Application of (b) leads to a global set of equations

$$\underset{\sim}{K} \underset{\sim}{u} + \underset{\sim}{F}_{0} = 0$$

in which

$$K_{ij} = \sum k_{ij}^{e} .$$

Similar applications are well known in electrical, hydraulic and other branches of engineering.

Object of Finite Element Method

The object of F.E.M. is to replace the infinite degree of freedom system in continuum applications by a finite system exhibiting the same basis as discrete analysis. This indeed is the definition of the F.E.M. process to which many different approaches lead.

Development

Stage I Intuitive Introduction

Early work Hrenikoff 1941 [1] and McHenry [2] replace structural continuum by equivalent bar assembly.

Argyris 1955 [3], Turner et al. 1956 [4] isolate continuum intuitively by assuming "reasonable behaviour pattern" for element.

Stage II Virtual Work and Energy Statement

Identification of process with virtual work or energy statements of structural mechanics.
Clough [5], Veubeke [6], Melosh [7], Zienkiewicz [8] and many others (1960-1965).
More formal requirements of application introduced.

Stage III Generalization and Recognition of Wider Mathematical Roots

Zienkiewicz and Cheung 1965 [9], Wilson and Nickell 1966 [10] and others show that F. E. M. can be applied to any mathematical problem in which a variational functional exists.

More recently it is shown that well known classical procedures known as "weighted residual methods" and which include

(a) Collocation

(b) Galerkin method (or Kantorovitch method)

(c) Least square approximation processes etc.

can all lead to FEM basis if piecewise continuous approximating functions are used, in a mathematical problem (each function defined uniquely within an element).

This allows a very wide expansion of process. Original work of Courant 1943 [11] and Prager and Synge 1947 [12] recognized as precursors of FEM.

Finite difference methods which originally appear to be a different process begin to be formulated on variational basis in 1960's and can be identified in FEM methodology.

Comment While mathematical bases of FEM are now established - and indeed a unified methodology now means that different mathematical techniques can be applied within same type of computer programs - the role of engineering (physical) intuition remains important.

Violation of strictly mathematical continuity conditions and relaxation of constraints guided by common sense often lead to dramatic improvement of approximation only some of which can be justified "a posteriori" at present.

Virtual Work as Basis for FEM Formulation of Solid Mechanics Problems. General

Let $\underset{\sim}{\sigma}$ be a stress system with corresponding body forces $\underset{\sim}{p}$ and surface tractions $\underset{\sim}{t}$.

Let $\underset{\sim}{\varepsilon}$ be a strain system with corresponding displacements $\underset{\sim}{f}$ in the interior and on the boundary.

Virtual work principle states that

(a) if $\underset{\sim}{\sigma}$, $\underset{\sim}{p}$ and $\underset{\sim}{t}$ are in equilibrium then the sum of internal and external work done during any virtual, compatible displacement ($\delta\underset{\sim}{\varepsilon}$ and $\delta\underset{\sim}{f}$) will be zero.

i.e.

$$\int_\Omega \delta\underset{\sim}{\varepsilon}^T \underset{\sim}{\sigma}\, d\Omega - \int_\Omega \delta\underset{\sim}{f}^T \underset{\sim}{p}\, d\Omega - \int_\Gamma \delta\underset{\sim}{f}^T \underset{\sim}{t}\, d\Gamma = 0$$

Ω being the domain of the solid and Γ its boundary.

The reverse statement is also true

(b) if $\underset{\sim}{\varepsilon}$ and $\underset{\sim}{f}$ are compatible strains and displacements, then the internal and external work done by applying these strains and displacements to any virtual and equilibrating system of forces and stresses ($\delta\underset{\sim}{p}$, $\delta\underset{\sim}{t}$ and $\underset{\sim}{\sigma}$) will be zero. i.e.

$$\int_\Omega \delta\underset{\sim}{\sigma}\, \underset{\sim}{\varepsilon}\, d\Omega - \int_\Omega \delta\underset{\sim}{p}\, \underset{\sim}{f}\, d\Omega - \int_\Omega \delta\underset{\sim}{t}\, \underset{\sim}{f}\, d\Gamma = 0 .$$

Both statements are well known for their application in structural mechanics.

By the first if we express the stresses (or their resultants) in terms of a set of parameters defining completely the displacement pattern; equilibrium relation can be obtained and the displacement parameters determined.

By the second, if we express the strains and displacements in terms of a set of parameters defining an equilibrating set of stresses and forces the set of stresses satisfying compatibility conditions is found. This has well known uses in conventional structural analysis ('dummy' load processes etc.).

If in the first case the parameters defining displacement do not permit the fully equilibrating position to be reached, the virtual work statement will ensure approximate equilibrium. Conversely if the second approach is used approximate compatibility will be achieved.

If the symbol δ used above is considered as a variation it is easy to show that the first approximation corresponds to well known principle of minimizing total potential energy, while second approximates to minimization of total complementary energy.

To derive elements we need not be restricted by the variational statements, if these exist same results can be obtained but wider possibilities exist.

Displacement Formulation

Let us define a field of displacement

$$\underset{\sim}{f} = \sum \underset{\sim}{N}_i \underset{\sim}{u}_i$$

with shape functions N_i defined piecewise by element. Compatible strains are defined as

$$\underset{\sim}{\varepsilon} = \sum \underset{\sim}{B}_i \underset{\sim}{u}_i$$

Constitutive law defines stresses as

$$\underset{\sim}{\sigma} = \underset{\sim}{\sigma}\left(\underset{\sim}{\varepsilon}\right)$$

Taking now a compatible virtual displacement

$$\delta \underset{\sim}{f} = \sum \underset{\sim}{W}_i \delta \underset{\sim}{u}_i$$

with strains

$$\delta \underset{\sim}{\varepsilon} = \sum \hat{\underset{\sim}{B}}_i \delta \underset{\sim}{u}_i$$

we have

$$\delta u_i^T \left(\int_\Omega \hat{\underset{\sim}{B}}_i^T \underset{\sim}{\sigma} \, d\Omega - \int_\Omega \underset{\sim}{W}_i^T p \, d\Omega - \int_\Gamma \underset{\sim}{W}_i^T \underset{\sim}{t} \, d\Gamma \right) = 0.$$

The quantity in brackets $= 0$ as δu_i arbitrary and immediately equation system for determining $\underset{\sim}{u}_i$ available.
If $\underset{\sim}{W}_i = \underset{\sim}{N}_i$ we have $\hat{\underset{\sim}{B}}_i = \bar{\underset{\sim}{B}}_i$ and method corresponds to

Galerkin process and is identical to minimizing total potential energy in conservative systems. Generally thus we have set of equations

$$\int_\Omega \bar{B}_i^T \sigma \, d\Omega - \int_\Omega N_i^T p \, d\Omega - \int_\Omega N_i^T t \, d\Gamma = 0$$

This with constitutive law allows solution to be reached. Basic FEM form exists if

$$\int_\Omega - d\Omega = \Sigma \int_{\Omega^l} - d\Omega$$

and contribution can be found element by element and summed (condition (b) basic for discrete system, see page 5).
So that integrals can be determined continuity conditions have to be imposed on N_i which depend on nature of B . Special example here is case of small elastic displacements with linear constitutive law

Now

$$\bar{B}_i = B_i$$

$$\sigma = D(\varepsilon - \varepsilon_0) + \sigma_0$$

where ε_0 and σ_0 initial strain and stress.
With body forces given by

$$p = p_0 - \mu \frac{\partial f}{\partial t} - \rho \frac{\partial f}{\partial t^2}$$

to account for viscous (damping) and inertia we have a set of equations

$$\underset{\sim}{K}\,\underset{\sim}{u} + \underset{\sim}{C}\,\frac{\partial}{\partial t}\underset{\sim}{u} + \underset{\sim}{M}\,\frac{\partial^2}{\partial t^2}\underset{\sim}{u} + \underset{\sim}{F}_0 = 0$$

with

$$\underset{\sim}{K}_{ij} = \int_\Omega \underset{\sim}{B}_i\,\underset{\sim}{D}\,\underset{\sim}{B}_j\,d\Omega \quad \text{(stiffness matrix)}$$

$$\underset{\sim}{C}_{ij} = \int_\Omega \underset{\sim}{N}_i\,\mu\,\underset{\sim}{N}_j\,d\Omega \quad \text{(damping matrix)}$$

$$\underset{\sim}{M}_{ij} = \int_\Omega \underset{\sim}{N}_i\,\varrho\,\underset{\sim}{N}_j\,d\Omega \quad \text{(mass matrix)}$$

$$\underset{\sim}{F}_{0i} = -\int_\Omega \underset{\sim}{N}_i\,\underset{\sim}{p}_0\,d\Omega - \int_\Omega \underset{\sim}{B}_i\,\underset{\sim}{D}\,\underset{\sim}{\varepsilon}_0\,d\Omega + \int_\Omega \underset{\sim}{B}_i\,\underset{\sim}{\sigma}_0\,d\Omega - \int_\Omega \underset{\sim}{N}_i\,\underset{\sim}{t}\,d\Gamma$$

well known standard forms.

Equilibrium Formulation

Now we shall assume that system of stresses can be defined which is in equilibrium with external forces. Piecewise definition in terms of a series of parameters is presumed

$$\underset{\sim}{\sigma} = \Sigma\,\underset{\sim}{Q}_i\,\underset{\sim}{a}_i$$

Constitutive law defines strains

$$\underset{\sim}{\varepsilon} = \underset{\sim}{\varepsilon}(\underset{\sim}{\sigma})$$

Taking now a virtual equilibrating stress change

$$\delta \underset{\sim}{\sigma} = \Sigma \underset{\sim}{Q}_i \, \delta \underset{\sim}{a}_i$$

($W_i = Q_i$ put in here immediately for clarity).

The corresponding virtual body forces and boundary traction are

$$\delta \underset{\sim}{p} = \Sigma A_i \, \delta \underset{\sim}{a}_i \qquad \delta t = \Sigma T_i \, \delta a_i$$

and by virtual work we have (immediately generalising for any $\delta \underset{\sim}{a}_i$) a set of equations

$$\int_\Omega \underset{\sim}{Q}_i^T \underset{\sim}{\varepsilon} \, d\Omega - \int_\Omega \underset{\sim}{A}_i^T \underset{\sim}{f} \, d\Omega - \int_\Omega \underset{\sim}{T}_i^T \underset{\sim}{f} \, d\Gamma = 0 .$$

With the constitutive law inserted it appears necessary to express $\underset{\sim}{f}$ in terms of $\underset{\sim}{\sigma}$ before useful results are obtained. It is usual to choose the parameters $\underset{\sim}{a}_i$ so that $\underset{\sim}{p}$ and prescribed boundary tractions are automatically equilibrated.

In such a case $\underset{\sim}{A}_i \equiv 0$ and $\underset{\sim}{T}_i \equiv 0$ on the portion of boundary where tractions known. Second integral disappears and last is determined where displacements known.

Equilibrium models are clearly more difficult to obtain than displacement ones.

Use of stress function is advantageous (Veubeke and Zienkiewicz [13] .

Mixed Formulations

If both displacements and stresses are prescribed in a manner which neither satisfies continuity nor equilibrium completely both sets of virtual work statements will have to be simultaneously satisfied.
Such approaches lead to same answers as mixed variational theorems (Hellinger-Reissner). Hybrid elements Pian [14; 18] are derived from such a basis.

Variational Form

The virtual work approach shown degenerates in a special case of $\underset{\sim}{W}_i = \underset{\sim}{N}_i$ to the same statement as minimization of total potential energy (in the displacement approach).

Quite generally if we have in many mathematical problems a functional

$$\mathcal{X}(\underset{\sim}{\Phi}) \longrightarrow \text{minimized}$$

where $\mathcal{X}$ is given by some domain and/or boundary integral.

$$\underset{\sim}{\Phi} = \sum \underset{\sim}{N}_i \underset{\sim}{a}_i$$

piecewise defined for approximate minimization leads to

$$\frac{\partial \mathcal{X}}{\partial \underset{\sim}{a}_i} = \sum \frac{\partial \mathcal{X}^e}{\partial \underset{\sim}{a}_i}$$

which is an essential FEM form.
If functional quadratic a linear equation system results.

Functionals sometimes available from physical basis (solid mechanics) or by mathematical reasoning if minimization leads to differential equation and boundary conditions of problem

$$\underset{\sim}{L}(\underset{\sim}{\Phi}) - \underset{\sim}{f} = 0 \qquad \text{in} \quad \Omega$$

and

$$\underset{\sim}{C}(\underset{\sim}{\Phi}) - \underset{\sim}{g} = 0 \qquad \text{on} \quad \Gamma$$

(Euler equations of the functional)
Advantage of variational forms is bound of $\mathcal{X}$, valuable if $\mathcal{X}$ has physical meaning.
Disadvantage that only for certain problems the direct path from differential to variation forms exists.

Weighted Residual Formulation [16-19]

If a differential equation governing $\underset{\sim}{\Phi}$ is

$$\underset{\sim}{A}(\underset{\sim}{\Phi}) \equiv \underset{\sim}{L}(\underset{\sim}{\Phi}) - \underset{\sim}{f} = 0 \quad \text{in} \quad \Omega$$

and the boundary condition is

$$\underset{\sim}{B}(\underset{\sim}{\Phi}) \equiv \underset{\sim}{C}(\underset{\sim}{\Phi}) - \underset{\sim}{g} = 0 \quad \text{on} \quad \Gamma$$

we can approximate directly

$$\bar{\underset{\sim}{\Phi}} = \sum \underset{\sim}{N}_i \underset{\sim}{a}_i .$$

Unless equation satisfied by chance exactly

$$\underset{\sim}{A}(\bar{\underset{\sim}{\Phi}}) = R_\Omega \neq 0 \quad \text{in} \quad \Omega$$

$$\underset{\sim}{B}(\bar{\underset{\sim}{\Phi}}) = R_\Gamma \neq 0 \quad \text{on} \quad \Gamma$$

R - "residual"

To achieve an approximation we can find parameters $\underset{\sim}{a}_i$ in such a way that residual is reduced to zero in the mean. If $\underset{\sim}{W}_j$, $\bar{W}_j$ represents weighting functions (j same number as that of unknown parameters $\underset{\sim}{a}_i$) sufficient equations can be obtained by orthogonality

$$\int_\Omega \underset{\sim}{W}_j^T \underset{\sim}{R}_\Omega \, d\Omega + \int_\Gamma \bar{\underset{\sim}{W}}_j^T \underset{\sim}{R}_\Gamma \, d\Omega = 0 .$$

Special Forms - Point Collocation

$$W_j = \delta_{ij} \quad \text{when} \quad x = x_j, \; y = y_j, \; z = z_j$$

$$W_j = 0 \quad \text{elsewhere}$$

$$\delta_{ij} - \text{Dirac delta}$$

Subdomain Collocation

$$W_j = 1 \quad \text{in region} \quad \Omega_j$$

$$W_j = 0 \quad \text{elsewhere}$$

Galerkin Process

$$\underset{\sim}{W}_j = \underset{\sim}{N}_j \; , \quad \bar{\underset{\sim}{W}}_j = -\underset{\sim}{N}_j$$

Note that discretization can be partial only when assuming approximation for $\Phi(x,y,z)$ expressed as

$$\bar{\underset{\sim}{\Phi}} = \Sigma \, \underset{\sim}{N}_i(x,y)\underset{\sim}{a}_i(z)$$

leaves discretized problem still as differential relationship but of reduced kind (Kantorovich).

Weighted residual processes are ideally suited to FEM as integrals involved again with property

$$\int_\Omega - d\Omega = \Sigma \int_\Omega - d\Omega$$

providing certain continuity observed.

Least Square Approximation

An alternative is to minimize the square of R with respect to $\underset{\sim}{a}_i$.

Approach is 'variational'

$$\mathcal{X} \rightarrow \min$$

$$\mathcal{X} = \int_\Omega R_\Omega^2 \, d\Omega + \int_\Gamma R_\Gamma^2 \, d\Gamma$$

$$\frac{\partial \mathcal{X}}{\partial a_i} = \int_\Omega 2 R_\Omega \frac{\partial R_\Omega}{\partial a_i} \, d\Omega + \int_\Gamma 2 R_\Gamma \frac{\partial R_\Gamma}{\partial a_i} \, d\Gamma$$

when a_i is a scalar, however more generally

$$\mathcal{X} = \int_\Omega \underset{\sim}{R}_\Omega^T \underset{\sim}{R}_\Omega \, d\Omega + \int \underset{\sim}{R}_\Gamma^T \underset{\sim}{R}_\Gamma \, d\Omega$$

leads to

$$\frac{\partial \mathcal{X}}{\partial \underset{\sim}{a}_i} = \int_\Omega 2 \underset{\sim}{R}_\Omega^T \frac{\partial \underset{\sim}{R}_\Omega}{\partial \underset{\sim}{a}_i} \, d\Omega + \int_\Gamma 2 \underset{\sim}{R}_\Gamma^T \frac{\partial R_\Gamma}{\partial \underset{\sim}{a}_i} \, d\Gamma$$

for vector quantities.

The minimization of the integrated square of the error leads to another form of weighted residual class with

$$\underset{\sim}{W}_i = \frac{\partial R_\Omega}{\partial \underset{\sim}{a}_i} \quad \text{in} \quad \Omega$$

$$\bar{\underset{\sim}{W}}_i = \frac{\partial R_\Gamma}{\partial \underset{\sim}{a}_i} \quad \text{on} \quad \Gamma$$

Galerkin Process

The various possibilities of approximation by weighting while offering diverse formulations suffer from one disadvantage against 'standard' variational formulation. To ensure integrability of

$$\int_\Omega \underset{\sim}{W}_i^T \underset{\sim}{R}_\Omega \, d\Omega$$

the continuity required of $\underset{\sim}{N}_i$ is of higher order than in variational form.

Weighting functions have however no continuity requirement.

Integration by parts will generally help and reduce continuity requirement on $\underset{\sim}{N}_i$, increasing these on $\underset{\sim}{W}_i$

In Galerkin process

$$\underset{\sim}{N}_i = \underset{\sim}{W}_i$$

and integration by parts is particularly desirable to reduce absolutely the continuity requirement.

Example of Applications of the various Processes

Problem:

$$\frac{\partial}{\partial x}\left(k\frac{\partial \Phi}{\partial x}\right) + \frac{\partial}{\partial y}\left(k\frac{\partial \Phi}{\partial y}\right) + Q = 0$$

in Ω

with

$$k\frac{\partial\Phi}{\partial h} - q = 0 \qquad \text{on} \quad \Gamma_1$$

$$\Phi - \Phi_B = 0 \qquad \text{on} \quad \Gamma_2$$

I. Variational

equivalent to

$$\text{minimum} \quad \mathcal{X} = \int_\Omega \frac{k}{2}\left[\left(\frac{\partial\Phi}{\partial x}\right)^2 + \left(\frac{\partial\Phi}{\partial y}\right)^2 - Q\cdot\Phi\right] d\Omega - \int_\Gamma q\,\Phi\, d\Gamma$$

subject to

$$\Phi = \Phi_B \qquad \text{on} \quad \Gamma_2$$

Take

$$\Phi = \sum N_i\, a_i$$

with some a_i determined to satisfy $\Phi = \Phi_B$ on Γ_2 and these parameters excluded later from minimization. We have

$$\frac{\partial\mathcal{X}}{\partial a_i} = \int_\Omega \left(k\,\frac{\partial\Phi}{\partial x}\cdot\frac{\partial N_i}{\partial x} + k\,\frac{\partial\Phi}{\partial y}\cdot\frac{\partial W_i}{\partial y} - Q\,N_i\right) d\Omega - \int_{\Gamma_1} q\,N_i\, d\Gamma = 0$$

Leading to

$$\underset{\sim}{H}\,\underset{\sim}{a} + \underset{\sim}{F}_0 = 0$$

$$h_{ji}^{e} = \int_{\Omega^e} k \left(\frac{\partial N_j}{\partial x} \cdot \frac{\partial N_i}{\partial z} + \frac{\partial N_j}{\partial y} \cdot \frac{\partial N_i}{\partial y} \right) d\Omega$$

$$F_{0j}^{e} = - \int_{\Omega^e} Q\, N_j \, d\Omega - \int_{\Gamma_1^e} q\, N_j \, d\Gamma$$

(Symmetry and Continuity needed)

II. Collocation (subdomain)

$W_j = 1$ in a jth part of each element

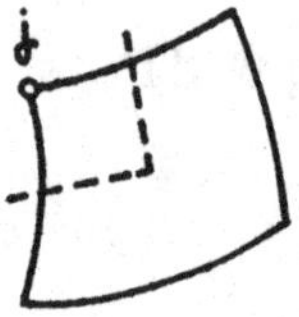

$$\int_{\Omega^e} W_j \left[\Sigma \left(\frac{\partial}{\partial x} k \frac{\partial N_i}{\partial z} + \frac{\partial}{\partial y} k \frac{\partial N_i}{\partial y} \right) a_i + Q \right] d\Omega -$$

$$- \int_{\Gamma_1^e} \bar{W}_j \left(\Sigma k \frac{\partial N_i}{\partial n} a_i - q \right) d\Gamma$$

or in FEM context

$$h_{ji}^{e} = \int_{\Omega^e} W_j \left(\frac{\partial}{\partial x} k \frac{\partial N_i}{\partial x} + \frac{\partial}{\partial y} k \frac{\partial W_i}{\partial y} \right) d\Omega - \int_{\Gamma_1^e} \bar{W}_j \frac{\partial N_i}{\partial n} d\Omega$$

$$F_{0j}^{e} = \int_{\Omega} Q\, W_j \, d\Omega - \int_{\Gamma} q\, \bar{W}_j \, d\Gamma$$

H now non symmetric

N_j needs C_1 continuity

(second integral on $\Gamma_2 = 0$ as $W_j = 0$ there).

III. Galerkin

All steps as above but we can integrate by parts. Total integral H_{ij} can be written now as

$$\int_{\Omega} N_j \left[\frac{\partial}{\partial x} k \frac{\partial N_i}{\partial z} + \frac{\partial}{\partial y} k \frac{\partial N_i}{\partial y} \right] d\Omega - \int_{\Gamma} N_j k \frac{\partial N_i}{\partial x} d\Gamma \equiv$$

$$\equiv - \int_{\Omega} k \left(\frac{\partial N_j}{\partial z} \cdot \frac{\partial N_i}{\partial x} + \frac{\partial N_j}{\partial y} \cdot \frac{\partial N_i}{\partial y} \right) d\Omega + \int_{\Gamma_1} N_j$$

and we therefore obtain identical expression to those found variationally (observe inclusion of boundary conditions in generalized Galerkin process).

IV. Least Squares

$$\frac{\partial R_{\Omega}}{\partial a_i} = \frac{\partial}{\partial x} \left(k \frac{\partial N_i}{\partial z} \right) + \frac{\partial}{\partial y} \left(k \frac{\partial N_i}{\partial y} \right)$$

$$\frac{\partial R_{\Gamma}}{\partial a_i} = \frac{\partial N_i}{\partial n} \quad \text{on} \quad \Gamma_1 = 0 \quad \text{on} \quad \Gamma_2$$

hence following it

$$h_{ji}^{e} = \int_{\Omega} \left[\frac{\partial}{\partial x} \left(k \frac{\partial N_j}{\partial z} \right) + \frac{\partial}{\partial y} \left(k \frac{\partial N_j}{\partial y} \right) \right] \left[\frac{\partial}{\partial z} \left(k \frac{\partial N_i}{\partial x} \right) + \frac{\partial}{\partial y} \left(k \frac{\partial N_i}{\partial y} \right) \right] d\Omega -$$

$$- \int_{\Gamma_1} \frac{\partial N_j}{\partial n} k \frac{\partial N_i}{\partial n} \, d\Gamma$$

$$F_{0j}^{e} = \int_{\Omega^e} Q \left[\frac{\partial}{\partial x} k \frac{\partial N_j}{\partial z} + \frac{\partial}{\partial y} k \frac{\partial N_j}{\partial y} \right] d\Omega + \int_{\Gamma_1} q \frac{\partial N_j}{\partial n} d\Gamma$$

H now symmetric but N_j still needs C_1 continuity.

Final Remark on Various Discretisation Processes

Variational form and Galerkin give same answer - both result in symmetric matrices and lowest continuity requirement. Collocation and least squares need higher continuity. Variation process difficult to apply in non linear case or if time variable enters. Consider non linear diffusion equation for which variational principle is not readily available

$$\frac{\partial}{\partial x}\left(k \frac{\partial \Phi}{\partial x}\right) + \frac{\partial}{\partial y}\left(k \frac{\partial \Phi}{\partial y}\right) + Q + c \frac{\partial \Phi}{\partial t} = 0$$

in Ω

with
$$\Phi = \Phi_B(t) \quad \text{on } \Gamma_2$$

$$k \frac{\partial \Phi}{\partial n} = q(t) \quad \text{on } \Gamma_1$$

Nonlinearity via

$$k = k(\Phi), \quad c = c(\Phi), \quad Q = Q(\Phi)$$

By Galerkin we take

$$\Phi = \sum N_i a_i \qquad \begin{aligned} N_i &= N_i(x, y) \\ a_i &= a_i(t) \end{aligned}$$

and as before

$$\underset{\sim}{H} \underset{\sim}{a} + \underset{\sim}{C} \frac{d}{dt} \underset{\sim}{a} + \underset{\sim}{F}_0 = 0$$

with

$$h_{ji}^e = \int_{\Omega^e} k \left(\frac{\partial N_i}{\partial x} \cdot \frac{\partial N_j}{\partial x} + \frac{\partial N_i}{\partial y} \cdot \frac{\partial N_j}{\partial y} \right) d\Omega$$

$$C_{ji} = \int_{\Omega^e} N_i \, c \, N_j \, d\Omega$$

$$F_{0i}^e = - \int_{\Omega^e} Q \, N_i \, d\Omega - \int_{\Gamma_1^e} g \, N_i \, d\Gamma$$

and a discretised form readily obtained.

A Non-Linear but Discretised System - Further Example

Two dimensional (plane) stress analysis approached directly by Galerkin method. Displacement formulae

$$\underset{\sim}{f} = \left\{ \begin{array}{c} u \\ v \end{array} \right\} \longrightarrow \underset{\sim}{\varepsilon} = \left\{ \begin{array}{c} \varepsilon_x \\ \varepsilon_y \\ \gamma_{xy} \end{array} \right\} \longrightarrow \underset{\sim}{\sigma} = \left\{ \begin{array}{c} \sigma_x \\ \sigma_y \\ \tau_{xy} \end{array} \right\} = \underset{\sim}{\sigma}(\underset{\sim}{f})$$

$$\underset{\sim}{f} = \Sigma \, N_i \left\{ \begin{array}{c} u_i \\ v_i \end{array} \right\}$$

We seek approximation satisfying equilibrium (i. e.)

in Ω

$$\frac{\partial \sigma_x}{\partial x} + \frac{\partial \tau_{xy}}{\partial y} + X = 0$$

$$\frac{\partial \sigma_y}{\partial y} + \frac{\partial \tau_{xy}}{\partial x} + Y = 0$$

$$\left\{ \begin{array}{c} X \\ Y \end{array} \right\} = \underset{\sim}{p}$$

on Γ_2

$$\sigma_x l_x - \tau_{xy} l_y = t_x$$

$$\sigma_y l_y - \tau_{xy} l_x = t_y$$

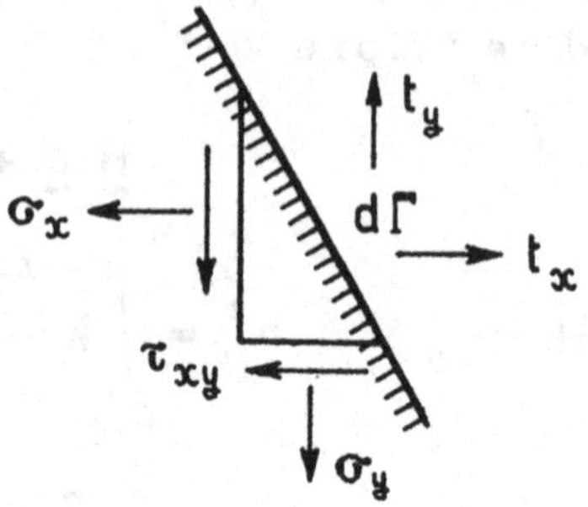

By Galerkin for i-th equation

$$\int_\Omega N_i\left(\frac{\partial\sigma_x}{\partial x} + \frac{\partial\tau_{xy}}{\partial y} + X\right)d\Omega - \int_{\Gamma_2} N_i\left(\sigma_x l_x - \tau_{xy} l_y - t_x\right)d\Gamma = 0$$

integrating by parts

$$-\int_\Omega\left(\sigma_x\frac{\partial N_i}{\partial x} + \tau_{xy}\frac{\partial N_i}{\partial y} - X N_i\right)d\Omega + \int_\Gamma \sigma_x l_x N_i\, dy + \tau_{xy} l_y N_i\, dx -$$

$$-\int_{\Gamma_2}\left(\sigma_x l_x - \tau_{xy} l_y - t_x\right) N_i\, d\Gamma = 0 .$$

Performing same operation in y direction and noting that by previous definition

$$\underset{\sim}{B}_i = \begin{bmatrix} \frac{\partial N_i}{\partial x}, & 0 \\ 0, & \frac{\partial N_i}{\partial y} \\ \frac{\partial N_i}{\partial y}, & \frac{\partial N_i}{\partial x} \end{bmatrix}$$

we have immediately (as $N_i = 0$ on Γ_1)

$$\int_\Omega \underset{\sim}{B}_i^T \underset{\sim}{\sigma}\, d\Omega - \int_\Omega \underset{\sim}{N}_i \underset{\sim}{p}\, d\Omega - \int \underset{\sim}{N}_i \underset{\sim}{t}\, d\Gamma = 0$$

the expression more directly obtained by virtual work previously.

Note this is again valid for any constitutive law - conservative or not.

Transformations as Galerkin for Discrete Equations

Galerkin (or variational processes) can be applied to transformation of discretized equation. This is useful to

(a) change geometric co-ordinates

(b) to further limit number of variables, e.g. eigenvalue economiser.

Consider we have discrete system

$$\underset{\sim}{H}\,\underset{\sim}{a} + \underset{\sim}{C}\,\dot{\underset{\sim}{a}} + \underset{\sim}{F} = 0 \qquad \frac{d\underset{\sim}{a}}{dt} \equiv a$$

and we further constrain the possible choice of parameters a_i so that

$$\underset{\sim}{a} = \underset{\sim}{T}\,\underset{\sim}{b}$$

$\underset{\sim}{T}$ plays role of weighting function hence orthogonality with $d\,\underset{\sim}{a}^T$ must be available. Thus

$$d\,\underset{\sim}{a}^T\left(\underset{\sim}{H}\,\underset{\sim}{a} + \underset{\sim}{C}\,\dot{\underset{\sim}{a}} + \underset{\sim}{F}\right) = 0$$

as
$$d\,\underset{\sim}{a} = T\,d\,\underset{\sim}{b}$$

or
$$\cancel{d\,\underset{\sim}{b}^T}\left(\underset{\sim}{T}^T\underset{\sim}{H}\,\underset{\sim}{T}\,\underset{\sim}{b} + \underset{\sim}{T}\,\underset{\sim}{C}\,\underset{\sim}{T}\,\dot{\underset{\sim}{b}} + \underset{\sim}{T}\,\underset{\sim}{F}\right) = 0$$

yield a new set of approximating equations. This is precisely the kind of transformation practical in structural mechanics but obtained by physical reasoning.

Shape Function of C_0 Continuity

(See Chapters 7 and 8 of ref. A)

1. For best results complete polynomial exponents needed.
2. 'Continuity' (C_0) ensured if along any side number of nodes = degree of expansion along $s - 1$.
3. Completeness satisfied if complete first order polynomial present.
4. Shape function N_i best derived 'by inspection' rather than inversion of polynomial.

Rectangular or Prism Elements

Always normalize co-ordinates

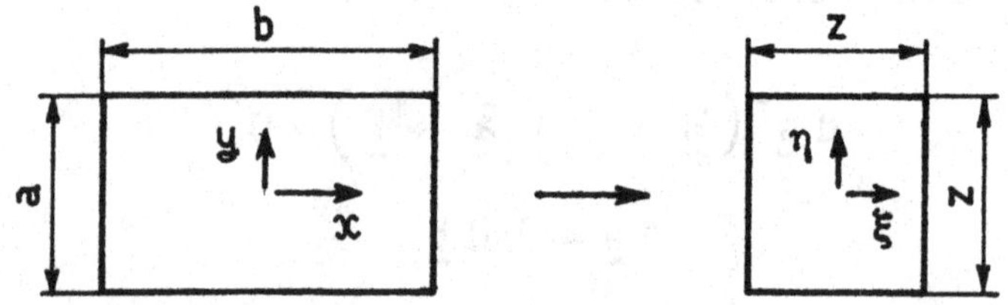

Transformation only requires constants.

Class A — Lagrangian

Simplest but least efficient

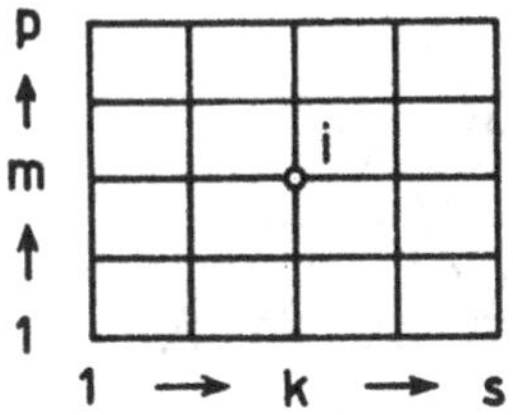

$$N_i = N_{km} = L_k^s L_m^p$$

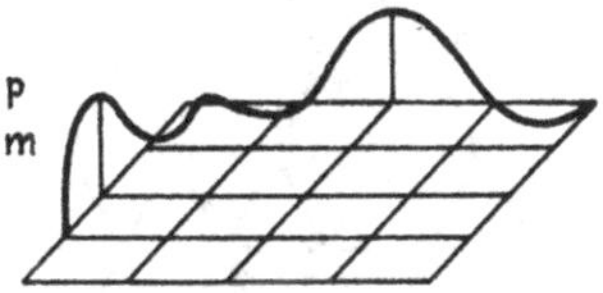

L_k^s Lagrangian polynomials

Class B — 'Serendipity' - Modes as Boundaries

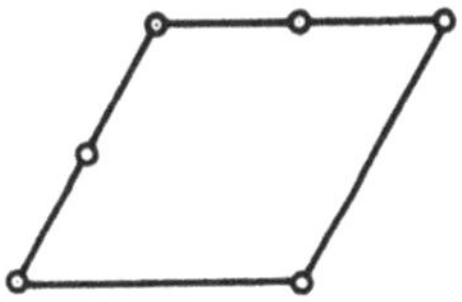

Combine

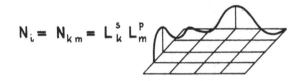

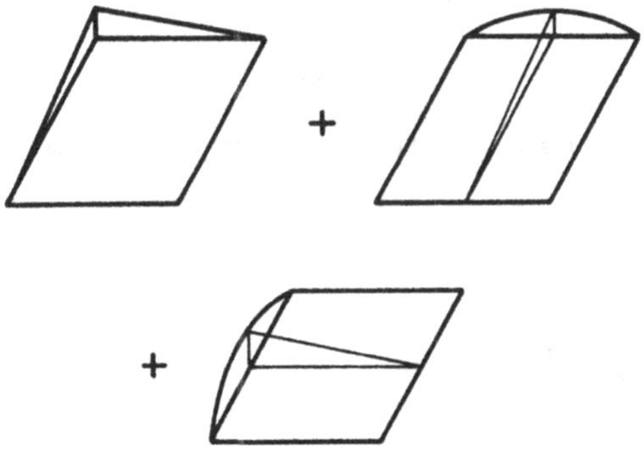

Very easy process of obtaining systematically elements with different numbers of nodes on boundaries (mixed elements).

Order of Polynomial Present (Pascal Triangle for polynomial terms)

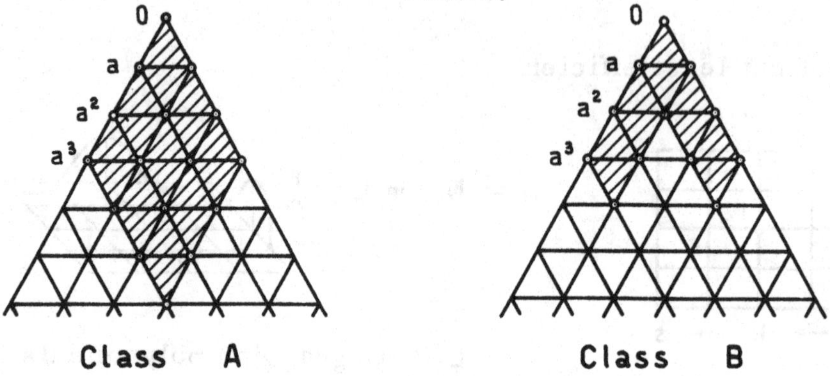

Class A Contains many 'extra' terms contributing little to accuracy.

Class B Up to third order OK, above needs addition of internal shape function.

Extension to third prism is obvious.

Triangles / Tetrahedra

Here the family can be made coincident with exact number of polynomial terms in Pascal triangle.

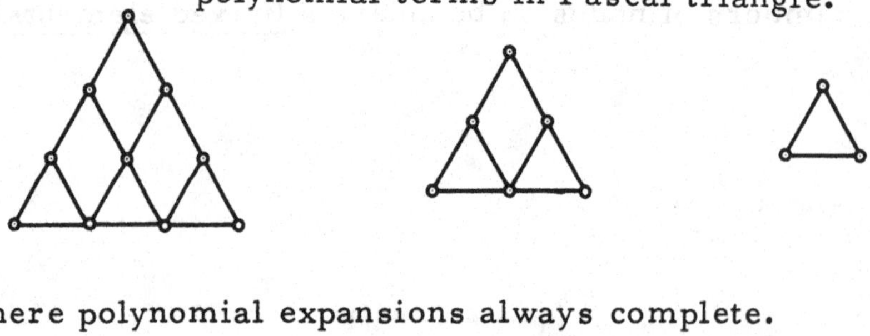

Note here polynomial expansions always complete.

For derivation of shape functions see Chapt. 7 (ref. A).

In terms of area/volume coordinates explicit expressions can be derived for any order of shape function.

Triangle can be used directly (and explicitly integrated).

Rectangle needs isoparametric transformation for practical use.

Nevertheless rectangle more popular as fewer elements needed to map a region.

Isoparametric Transformation

(See Ch. 8, ref. A - only very brief note here).

Region of rectangle (or prism) mapped taking

$$x = \Sigma N_i x_i$$

$$y = \Sigma N_i y_i$$

using $N_i(\xi, \eta)$ as in the interpolation of unknown

$$\Phi = \Sigma N_i a_i$$

This mapping is advantageous as

(a) ensures that complete first order terms in x, y present for Φ

(b) continuity of subdivision and Φ (C_0) assured between elements.

However,

(c) in distorted form available expression for Φ no longer polynomial in x, y .

(d) if distortion excessive mapping may not be 'one to one'.

To determine x, y derivatives of Φ we determine Jacobian matrix

$$\underset{\sim}{J} = \frac{\partial(x,y)}{\partial(\xi,\eta)} \equiv \begin{bmatrix} \frac{\partial x}{\partial \xi} , & \frac{\partial x}{\partial \eta} \\ \frac{\partial y}{\partial \xi} , & \frac{\partial y}{\partial \eta} \end{bmatrix}$$

$$\begin{Bmatrix} \frac{\partial}{\partial x} \\ \frac{\partial}{\partial y} \end{Bmatrix} = \underset{\sim}{J}^{T-1} \begin{Bmatrix} \frac{\partial}{\partial \xi} \\ \frac{\partial}{\partial \eta} \end{Bmatrix}$$

As

$$\frac{\partial x}{\partial \xi} = \sum \frac{\partial N_i}{\partial \xi} x_i \qquad \text{etc.}$$

J can be determined numerically.

Further vectors

$$\underset{\sim}{d\xi} = \begin{Bmatrix} \frac{dx}{d\xi} \\ \frac{dy}{d\xi} \\ \frac{dz}{d\xi} \end{Bmatrix} d\xi$$

etc. are tangent in the x, y, z space to ξ, η, ζ constant lines

Element of volume given as triple scalar produces

$$d(\text{vol}) = d\underset{\sim}{\xi} \cdot (d\underset{\sim}{\eta} \cdot d\underset{\sim}{y}) = \det|\mathfrak{J}|\, d\xi\, d\eta\, dy.$$

Numerical integration will always be needed and simple for determination of element matrices of type

$$\int_{\Omega} G\, dx\, dy\, dz$$

Note vectors $d\underset{\sim}{\xi}$ etc. form the columns of the $\mathfrak{J}$ matrix. If expansion is linear in any direction such as η, this vector will be constant and can be written as

$$d\underset{\sim}{\eta} = b\, \underset{\sim}{c}\, d\eta$$ where $\underset{\sim}{c}$ is

a unit vector and both b and c are only functions of two variables ξ and η and one can reduce calculation of $\mathfrak{J}^T$ to two variables and thus limit numerical integration to two functions. (or one in two dimensional problems).

Numerical Integration – Required Accuracy

In isoparametric elements (and often others) numerical integration advantageous (see Chap. 8 ref. A) e.g.

$$\int_{-1}^{1}\int_{-1}^{1} G(\xi,\eta)\, d\xi\, d\eta = \sum G(\xi_i,\eta_j)\cdot H_{ij}$$

H_{ji} - weighting factors.

The accuracy of results perhaps surprisingly does not increase with order of integration hence minimum order consistent with convergence needed.

For convergence element has to reproduce correct nodal forces as its size $\rightarrow 0$ and "stresses" become constant. (Note the same argument valid for any problem where $\underset{\sim}{B}$ depends on first derivatives). As

$$\underset{\sim}{F} = \int_{\Omega^e} \underset{\sim}{B}^T \underset{\sim}{\sigma} \, d\underset{\sim}{\Omega} \qquad \text{as } \sigma\text{-const}$$

all terms

$$\int_{\Omega^e} B_i^{km} \, d\Omega$$

should be correctly integrated or

$$\int \frac{\partial N_i}{\partial x} \, d\Omega \, , \quad \int \frac{\partial N_i}{\partial y} \, d\Omega$$

should be exact.

Noting that

$$d\Omega = \det. \left| J \right| d\xi \, d\eta$$

and that

$$\begin{Bmatrix} \dfrac{\partial N_i}{\partial x} \\ \dfrac{\partial N_i}{\partial y} \end{Bmatrix} = \underset{\sim}{J}^{-1} \begin{Bmatrix} \dfrac{\partial N_i}{\partial \xi} \\ \dfrac{\partial N_i}{\partial \eta} \end{Bmatrix}$$

shows that integrals of products such as

$$\int_{\Omega^e} \frac{\partial N_i}{\partial \eta} \cdot \frac{\partial N_j}{\partial \xi} \, d\xi \, d\eta$$

must be correctly integrated.

This is equivalent to saying that

$$\int_{\Omega} \det|\mathfrak{J}|\, d\xi\, d\eta$$

must be integrated exactly or that the volume integration must be correct for any distorted form.

Following general propositions can be made

(a) 'Inexactly' integrated elements will be softer than corresponding exactly integrated elements.

(b) 'Bounding' nature of approximation is lost.

(c) Generally better results will be obtained.

Dramatic improvement obtained with parabolic and linear element.

Wilson et al. [20] show improvement in linear quadrilaterals by selectively reduced integration, Zienkiewicz et al. [21] in parabolic, thin elements (see Chap. 14, ref. A).

Excessive reduction of integration may cause singularity greater than that of rigid body modes. For no singularity we could argue that Number of D.O.F.(°) = Number of independent equations supplied at Gauss points = Number of rigid body modes. For example consider a linear element reduced to

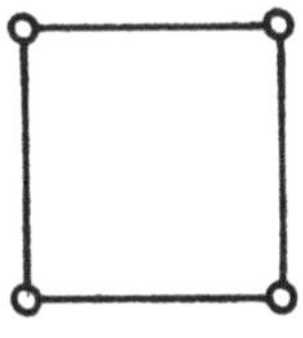

one integrating point. We have 8 D.O.F. and from the 3 stresses, 3 independent equations.

(°) D.O.F.= degree of freedom

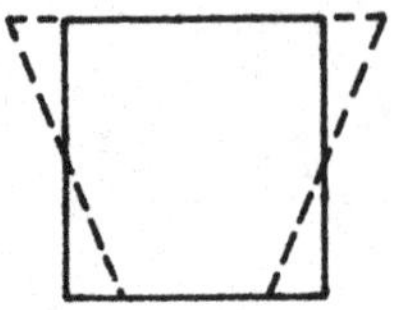

Difference $5 > 3$. Rigid body modes available. Hence element is singular. Such singular modes are illustrated.

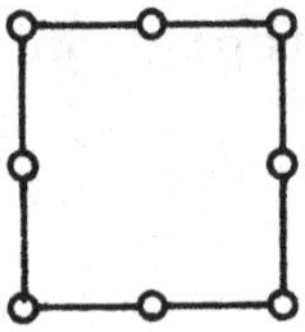

Parabolic element with 4 integration points and 16 D.O.F.

$3 \times 4 = 12$ independent equations

$16 - 12 = 4 > 3$ singularity again existing.

But on assembly of several element singularity may disappear. This happens in parabolic 2 or 3rd elements when more than 2 put together.

Note Singularity Conditions for Single Point Element

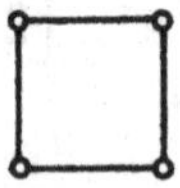

Single point integrated element

DOF $8 - 3 = 5 > 3$ singular

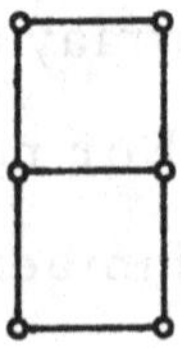

Assembly of two elements

DOF $12 - 2 \times 3 = 6 > 3$ singular

Consider an assembly $n \times m$ elements.

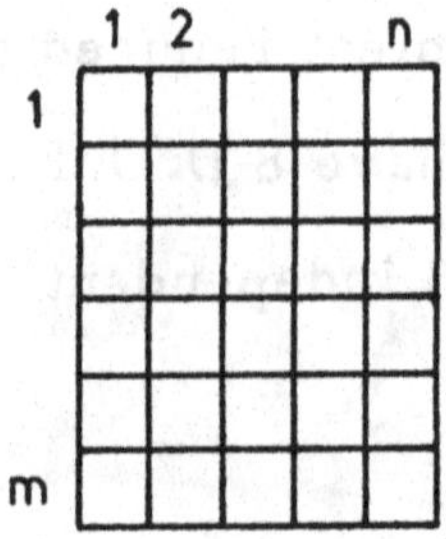

DOF $(n+1)(m+1) \times 2$

$- 3 \times n \times m$ eq. $=$

$= 2nm + 2n + 2m + 2 - 3nm$

$= 2(n+m) + 2 - nm < 3$

if $n = m$

$$N = 4n + 2 - n^2 = 2(n+1)^2 - 3n^2$$

$n = 1$	$N = 5 > 3$	singular
$n = 2$	$N = 8 + 2 - 4 = 6 > 3$	singular
$n = 3$	$N = 12 + 2 - 9 = 5 > 3$	singular
$n = 4$	$N = 16 + 2 - 16 = 2 < 3$	non singular

if more than 4 x 4 mesh used we are O.K.

If same applied to a 3D element

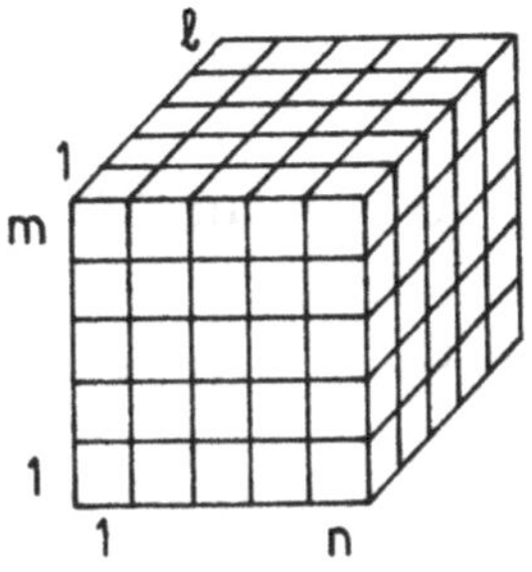

$$DOF - IND\ EQ < 6$$

$$DOF = (n+1)(m+1)(\ell+1)\cdot 3$$

$$IND\ EQ. = n\cdot m\cdot \ell \cdot 6 \quad (6\ \text{stress}).$$

Condition for no singularity

$$3(n+1)(m+1)(\ell+1) - 6\,nm\ell < 6$$

taking $n = m = \ell$

$$\bar{N} = 3(n+1)^3 - 6n^3$$

$n = 1$	$\bar{N} = 3.2^3 - 6 = 24 - 6 = 18 > 6$
$n = 2$	$= 3.3^3 - 6.8 = 33 > 6$
$n = 3$	$= 30 > 6$
$n = 4$	$= -9$ O.K.

assembly of 4 x 4 x 4 is O.K.

Finite Differences (Variational FOrm) – Relation with Incompletely Integrated Elements

In variationally conceived finite difference

(1) approximation to derivatives occurring in variational form made locally

(2) this approximation applied to a certain area and derivatives found

(3) Assembly following finite element scheme made.

Example Regular mesh $\nabla^2 \Phi - Q = 0$ in Ω

with $\frac{\partial \Phi}{\partial n} - q = 0$ on Γ_1

$\Phi = \Phi_2$ in Γ_2

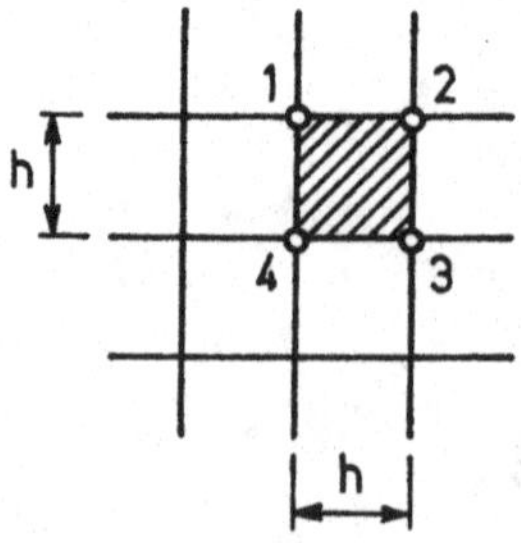

$$\chi = \int_\Omega \frac{1}{2} \left[\left(\frac{\partial \Phi}{\partial x}\right)^2 + \left(\frac{\partial \Phi}{\partial y}\right)^2 \right] dx\, dy - $$

$$- \int Q \cdot \Phi \, dx\, dy - \int q\, \Phi \, d\Gamma \, .$$

is mode minimum.

In each elementary region for interior point we approximate

$$\frac{\partial \Phi}{\partial x} = \left(\Phi_2 - \Phi_1 - \Phi_3 - \Phi_4\right)/2h$$

and similar for $\frac{\partial \Phi}{\partial y}$ and element contribution found.

This is exactly equivalent to reducing integration in a linear

FEM to one point.

(Assembled eq. is the usual Fin. diff. applied (°) to diagonal pattern but interweaving).

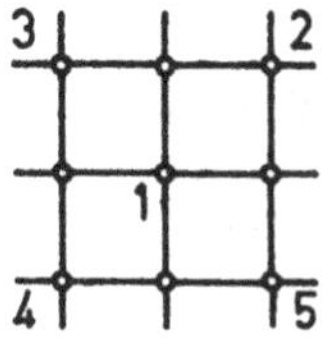

$$-4\,\Phi_1 + \Phi_2 + \Phi_3 + \Phi_4 + \Phi_5 + 2h^2Q$$

Other cells can be created as

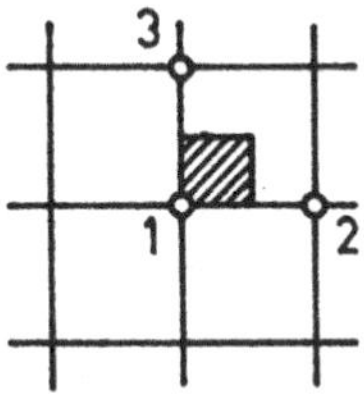

in this take

$$\frac{\partial\Phi}{\partial x} = (\Phi_2 - \Phi_1)/h$$

$$\frac{\partial\Phi}{\partial y} = (\Phi_3 - \Phi_1)/h$$

(vide Griffin and Kellog 1967 [23]).

In either case we have finite elements. The first in conventional, the second with external nodes.

A disadvantage of finite difference forms is regular mesh.

This can be overcome by using isoparametric form of coordinates, e.g. in ξ, η coordinates we can still use approximation

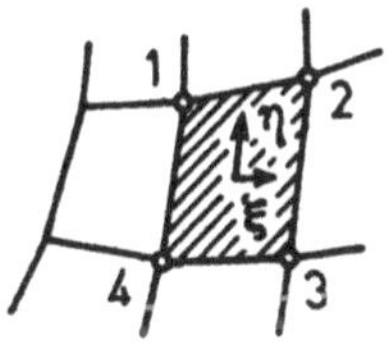

$$\frac{\partial\Phi}{\partial\xi} = (\Phi_2 - \Phi_1 - \Phi_3 - \Phi_4)/4$$

etc., and then use usual Jacobian transformation to obtain

$$\frac{\partial\Phi}{\partial x} \text{ etc.}$$

(°) Clearly if insufficient constraints applied to this problem equation can not be solved due to some singularity as in reduced integration elements.

In this case again we shall simply find the 'single point' integrated element.
Other possibilities are obvious and are being investigated but identity of finite differences and finite elements in some contexts should be noted.

Thick Shells and Plates as a "Degeneration" of Solids

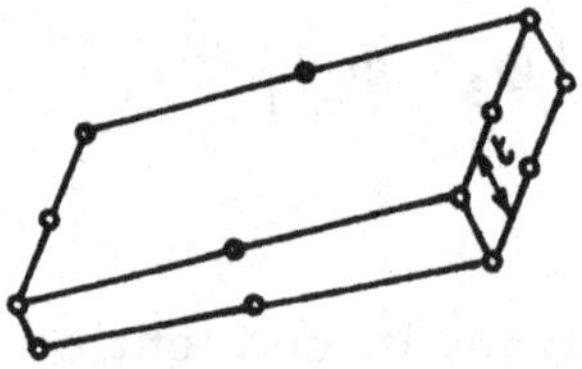

(See Ch. 14, ref. A)
3D element can be degenerated to thin shape and thus represent plate or shell. Kirchhoff assumption will not be invoked so shear deformation present. It seems natural to use an element with linear form in thickness.

Difficulties

(a) Possible ill-conditioning of equation as stiffness in lateral direction large compared with others.
(b) More degrees of freedom than necessary in shell (6 in place of 5).
(c) Linear variation of displacement across (ζ) means that σ_y is constant. This induces strains in shell plane if $\nu \neq 0$

which are not correct.

(d) Spurious shear may affect performance.

Solution of these difficulties involves

(a) disregarding relative displacement in lateral direction in strain determination

(b) assuming $\sigma_y = 0$

(c) behaviour can be improved by reduced integration

In general shell calculation certain transformation needed - process easily explained in flat plate case

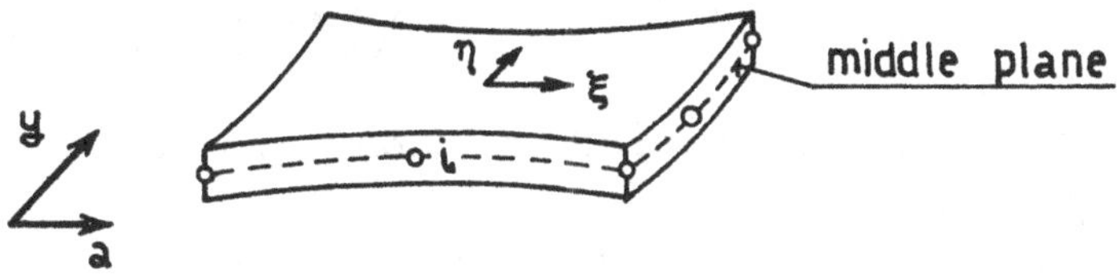

Assume

1) middle plane flat contains ξ and η

2) coordinate z is normal to middle plane

3) thickness interpolated

$$t = \sum N_i t_i \qquad N_i = N_i(\xi, \eta)$$

with i associated with middle plane node

6 DOF $\quad (u_i, v_i, w_i)^{top}$ and $(u_i, v_i, w_i)^{bot}$

replaced by

5 DOF $\quad u_i, v_i, w_i \quad$ in middle plane

Θ_{xi} rotation about y axis

Θ_{yi} rotation about x axis

displacements u, v, ω

are

$$\begin{Bmatrix} u \\ v \\ \omega \end{Bmatrix} = \begin{Bmatrix} u \\ v \\ \omega \end{Bmatrix}_{min} + z \begin{Bmatrix} \Theta_y \\ -\Theta_x \\ 0 \end{Bmatrix}$$

or assuming same interpolation for all variables

$$\begin{Bmatrix} u \\ v \\ \omega \end{Bmatrix} = \sum N_i \left(\begin{Bmatrix} u_i \\ v_i \\ \omega_i \end{Bmatrix} + z \begin{Bmatrix} \Theta_{xi} \\ -\Theta_{yi} \\ 0 \end{Bmatrix} \right)$$

strain to be considered will now involve

$$\begin{Bmatrix} \epsilon_x \\ \epsilon_y \\ \gamma_{xz} \\ \gamma_{yz} \\ \gamma_{xy} \end{Bmatrix} = \begin{bmatrix} \frac{\partial}{\partial x} & 0 & 0 \\ 0 & \frac{\partial}{\partial y} & 0 \\ \frac{\partial}{\partial z} & 0 & \frac{\partial}{\partial x} \\ 0 & \frac{\partial}{\partial z} & \frac{\partial}{\partial y} \\ \frac{\partial}{\partial y} & \frac{\partial}{\partial x} & 0 \end{bmatrix} \begin{Bmatrix} u \\ v \\ \omega \end{Bmatrix} = \sum \left[\hat{\underset{\sim}{B}}_i , \check{\underset{\sim}{B}}_i \right] \left\{ \delta_i \right\}$$

$$\delta_i = \begin{Bmatrix} u_i \\ v_i \\ \omega_i \\ \Theta_{xi} \\ \Theta_{yi} \end{Bmatrix} \qquad \hat{\underset{\sim}{B}} = \begin{bmatrix} \frac{\partial N_i}{\partial x} & 0 & 0 \\ 0 & \frac{\partial N_i}{\partial y} & 0 \\ 0 & 0 & \frac{\partial N_i}{\partial x} \\ 0 & 0 & \frac{\partial N_i}{\partial y} \\ \frac{\partial N_i}{\partial x} & \frac{\partial N_i}{\partial y} & 0 \end{bmatrix}$$

$$\check{\underset{\sim}{B}}_i = \begin{bmatrix} \frac{\partial N_i}{\partial x} z & 0 \\ 0 & -\frac{\partial N_i}{\partial y} z \\ N_i & 0 \\ 0 & N_i \\ \frac{\partial N_i}{\partial y} z & \frac{\partial N_i}{\partial x} z \end{bmatrix} = \begin{bmatrix} 0 & 0 \\ 0 & 0 \\ N_i & 0 \\ 0 & N_i \\ 0 & 0 \end{bmatrix} + z \begin{bmatrix} \frac{\partial N_i}{\partial x} & 0 \\ 0 & \frac{\partial N_i}{\partial y} \\ 0 & 0 \\ 0 & 0 \\ \frac{\partial N_i}{\partial y} & \frac{\partial N_i}{\partial x} \end{bmatrix}$$

The $\underset{\sim}{D}$ matrix is a 5 x 5 one omitting σ_2 but obtained by taking $\sigma_2 = 0$.

For isotropic material

$$\underset{\sim}{D} = \frac{E}{1-\nu^2} \begin{bmatrix} 1 & \nu & 0 & 0 & 0 \\ & 1 & 0 & 0 & 0 \\ & & \frac{1-\nu}{2k} & 0 & 0 \\ & s > m & & \frac{1-\nu}{2k} & \\ & & & & \frac{1-\nu}{2} \end{bmatrix}$$

($k = 1{,}2$ is taken into account for non uniform shear stress)

As $$N_i = N_i(\xi, \eta)$$

and $$x = \sum N_i x_i \qquad y = \sum N_i y_i$$

usual transformation in 2-dimensional mode

$$\begin{Bmatrix} \frac{\partial N_i}{\partial x} \\ \frac{\partial N_i}{\partial y} \end{Bmatrix} = \begin{vmatrix} \frac{\partial x}{\partial \xi} & \frac{\partial y}{\partial \xi} \\ \frac{\partial x}{\partial \eta} & \frac{\partial y}{\partial \eta} \end{vmatrix}^{-1} \begin{Bmatrix} \frac{\partial N_i}{\partial \xi} \\ \frac{\partial N_i}{\partial \eta} \end{Bmatrix}$$

element of volume

$$d\Omega = dy\, t\, d\xi\, d\eta\, \det|J|$$

Stiffness can be calculated in usual way

$$\underset{\sim}{k}_{ij} = \underset{\sim}{B}_i^T \underset{\sim}{D} \underset{\sim}{B}_j \, d\Omega$$

t obtained by interpolation $t = \Sigma N_i t_i$

Numerical integration is confined to ξ , η plane

$$\underset{\sim}{B}_i^T \underset{\sim}{D} \underset{\sim}{B}_j = \begin{bmatrix} \hat{B}_i^r \\ \check{B}_i^r \end{bmatrix} \underset{\sim}{D} \left[\hat{B}_j , \check{B}_j \right] .$$

Some terms contain y some y^2 . Integration with respect to y can be carried out explicitly.

Any integration to be carried out numerically across thickness can be suitably organised to save time.

Extension of the process to curved shells needs several transformations which are detailed elsewhere. Procedures essentially the same.

Porous Materials – Soil, Pock, Concrete – Some Common Characteristics

1. Materials - two phase (or more) (solid-fluid)
2. Solid 'skeleton' properties with zero pressure fundamental and assumption made that these are not influenced by presence of fluid (adhesion etc. ignored).
3. Fluid in the pores has very small shear stresses and can be described by 'p' pore pressure even in motion.

Problems fall into three classes:

<u>I. Long term - fully drained situation</u>

Here p pore pressure, distribution known explicitly (possibly determined from flow characteristics).

<u>II. Instantaneous loading - undrained situation</u>

Here p unknown and must be determined in solution. No fluid flow occurs between pores and material behaviour almost incompressible.

<u>III. Intermediate time - coupled settlement</u>

Here the flow problems and straining of skeleton are coupled. Flow problems influenced by

(a) rate of solid skeleton strain

(b) changes of permeability with strain.

Straining in turn influenced as in I and II by pore pressure.

We shall discuss problems in that order.

I. Long Term - Fully Drained

Total stress $\underset{\sim}{\sigma} = \underset{\sim}{\sigma}_e + \underset{\sim}{\sigma}_h$

$\underset{\sim}{\sigma}_e$ effective stress

$\underset{\sim}{\sigma}_h$ hydrostatic stress $= - \begin{Bmatrix} 1 \\ 1 \\ 1 \\ 0 \\ 0 \\ 0 \end{Bmatrix} p$

If $\underset{\sim}{\varepsilon}$ strain of skeleton

$$\underset{\sim}{\varepsilon} = \underset{\sim}{\varepsilon}(\underset{\sim}{\sigma}_e) \quad \text{or} \quad \underset{\sim}{\sigma}_e = \underset{\sim}{\sigma}_e(\underset{\sim}{\varepsilon})$$

i.e. strain of skeleton depends only in effective stress in first approximation.

(To this one can add a uniform volumetric strain of small magnitude

$$\underset{\sim}{\varepsilon}_h = \underset{\sim}{D}_h \underset{\sim}{\sigma}_h$$

if appreciable skeleton compressibility exists. This here neglected). Equilibrium equations are always valid in terms of total stress i.e. in displacement formulation we have the standard form

$$\int_\Omega \underset{\sim}{B}^T \underset{\sim}{\sigma} \, d\Omega + \underset{\sim}{F} = 0$$

in which $\underset{\sim}{F}$ determined in terms of total external loads. Dividing

$$\int_\Omega \underset{\sim}{B}^T \underset{\sim}{\sigma} \, d\Omega = \int_\Omega \underset{\sim}{B}^T \underset{\sim}{\sigma}_e \, d\Omega + \int_\Omega \underset{\sim}{B}^T \underset{\sim}{\sigma}_h \, d\Omega$$

leaves first term which can be determined from strains alone, second term which can be evaluated from pressure p which may be known a priori i.e.

$$\underset{\sim}{F}_h = \int_\Omega \underset{\sim}{B}^T \underset{\sim}{\sigma}_h \, d\Omega = -\int_\Omega B^T \begin{Bmatrix} 1 \\ 1 \\ 1 \\ 0 \\ 0 \\ 0 \end{Bmatrix} p \, d\Omega$$

i.e. in terms of effective stress we re-formulate the problem as

$$\int_\Omega \underset{\sim}{B}^T \underset{\sim}{\sigma}_e \, d\Omega + \underset{\sim}{\bar{F}} = 0$$

with

$$\underset{\sim}{\bar{F}} = \underset{\sim}{F} + \underset{\sim}{F}_h .$$

Standard form of analysis of linear or non-linear type can be used. The only difference is the pore pressure force which is evaluated in above.

A. Note

Any standard FEM elastic program containing input for thermal strain can be converted to deal with pore pressure. If material elastic then thermal "body forces" are given as

$$= \underset{\sim}{B}^T \underset{\sim}{D} \, \underset{\sim}{\varepsilon}_0 \, d\Omega \equiv \underset{\sim}{B}^T \underset{\sim}{D} \, \alpha \begin{Bmatrix} 1 \\ 1 \\ 1 \\ 0 \\ 0 \\ 0 \end{Bmatrix} T \, d\Omega$$

stress

$$\underset{\sim}{\sigma} = \underset{\sim}{D} \, \underset{\sim}{\varepsilon} - D\varepsilon_0 .$$

Body forces identical to those given by pressure if

$$D\alpha \begin{Bmatrix} 1 \\ 1 \\ 1 \\ 0 \\ 0 \\ 0 \end{Bmatrix} T \rightarrow - \begin{Bmatrix} 1 \\ 1 \\ 1 \\ 0 \\ 0 \\ 0 \end{Bmatrix} p$$

stress which is given is total stress (not effective)

To ensure identity

$$(D_{11} + D_{12} + D_{13})\,\alpha T = -p$$

$$(D_{21} + D_{22} + D_{23})\,\alpha T = -p$$

$$(D_{31} + D_{32} + D_{33})\,\alpha T = -p$$

or $T = p\ \alpha (D_{11}+D_{12}+D_{13})$ if material isotropic. Similar transformation possible if material anisotropic; now α_s will be directional.

B. Note

A well known alternative to above is to look at differential equations of equilibrium and note that if separation of effective and hydraulic stresses is made the effect is to put in body forces

$$\begin{Bmatrix} X \\ Y \\ Z \end{Bmatrix} = - \begin{Bmatrix} \frac{\partial}{\partial x} \\ \frac{\partial}{\partial y} \\ \frac{\partial}{\partial z} \end{Bmatrix} p$$

This is the formulation described in text (A) . Now again we reduce solution to

$$\int_\Omega \underset{\sim}{B}^T \underset{\sim}{\sigma}_e + \hat{\underset{\sim}{F}} = 0$$

But

$$\hat{\underset{\sim}{F}} = \check{\underset{\sim}{F}} + \hat{\underset{\sim}{F}}_h$$

where $\check{\underset{\sim}{F}}$ gives boundary forces in terms of effective terms (i. e. after subtracting pore pressure)

and

$$\hat{\underset{\sim}{F}}_h = \int_\Omega \underset{\sim}{N}^T \begin{Bmatrix} \frac{\partial}{\partial x} \\ \frac{\partial}{\partial y} \\ \frac{\partial}{\partial z} \end{Bmatrix} p \, d\Omega$$

By integration by parts it is easy to show that

(a) at internal nodes identical forces will result with previous results

(b) that at external nodes the difference of contribution of $\check{F}$ and $\hat{F}_h$ is such as makes forces there identical.

The method described (suggested by D. Naylor) is easier as p can, if wished, be discontinuous. Also boundary forces are more "obvious".

In deformation studies of engineering problems note that differences of original and final pressures gives active forces. E. g. in dam foundation differences of p_0 and p_t before and after construction important (vide O. C. Zienkiewicz in Ch. 7 Handbook of Applied Hydraulics 1968).

Example Deformation of a geological structure due to oil pressure depletion

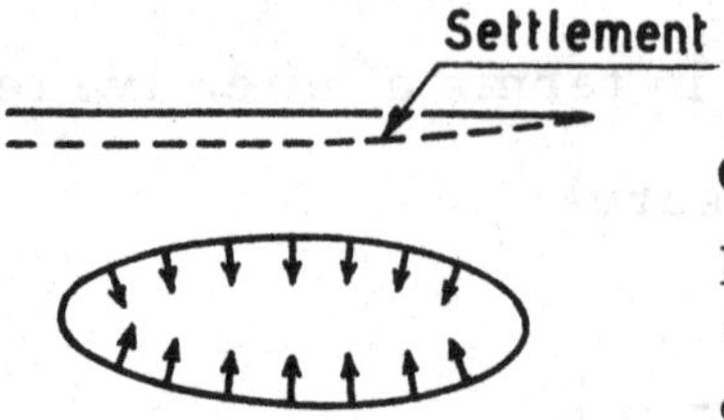

Oil bearing inclusion in which pressure decrease (uniform) caused by pumping gives a set of boundary forces at interface shown which results in a settlement.

II. Instantaneous Load - Undrained Situation

We observe that as no fluid escapes

(a) change of fluid volume = change of total volume (if skeleton particles are incompressible - again if skeleton compressible a correction made here).

(b) the total fluid pressure change p can be related to changes of fluid volume i.e.

$$\underset{\sim}{\sigma}_h = \underset{\sim}{\sigma}_h(\bar{\varepsilon}) \quad \text{or} \quad p = p(\bar{\varepsilon})$$

where $\bar{\varepsilon}$ = volume change = $\varepsilon_x + \varepsilon_y + \varepsilon_z$ =

$$= \begin{Bmatrix} 1 \\ 1 \\ 1 \\ 0 \\ 0 \\ 0 \end{Bmatrix}^T \underset{\sim}{\varepsilon}$$

If water only is considered compressibility is small and a linear relation exists

$$p = -\frac{1}{\eta} K_f \bar{\varepsilon}$$

where K_f - bulk modulus η -porosity.

If gas/fluid mixture fills the pores a more complex relation can easily be established. Quite generally

$$\underset{\sim}{\sigma}_h = \underset{\sim}{\sigma}_h(\underset{\sim}{\varepsilon}).$$

Now equilibrium given as before in total stress terms results in

$$\int_\Omega \underset{\sim}{B}^T \underset{\sim}{\sigma} \, d\Omega + \underset{\sim}{F} = 0$$

with

$$\underset{\sim}{\sigma} = \underset{\sim}{\sigma}_e(\underset{\sim}{\varepsilon}) + \underset{\sim}{\sigma}_h(\underset{\sim}{\varepsilon}) = \underset{\sim}{\sigma}(\underset{\sim}{\varepsilon})$$

and standard process used in solution whether material elastic or inelastic.

From strains determined $\underset{\sim}{\sigma}_e$ and $\underset{\sim}{\sigma}_h$ are separately given.

Note: In practice with saturated soil volume deformation at this stage small and linear elastic behaviour taken

Thus:

$$\underset{\sim}{\sigma}_e = \underset{\sim}{D}_e \underset{\sim}{\varepsilon}$$

$$\underset{\sim}{\sigma}_h = -\begin{Bmatrix} 1 \\ 1 \\ 1 \\ 0 \\ 0 \\ 0 \end{Bmatrix} p = \begin{Bmatrix} 1 \\ 1 \\ 1 \\ 0 \\ 0 \\ 0 \end{Bmatrix} \frac{K_f}{\eta} \begin{Bmatrix} 1 \\ 1 \\ 1 \\ 0 \\ 0 \\ 0 \end{Bmatrix}^T \underset{\sim}{\varepsilon}$$

cont.

This gives conveniently the total strain elastic matrix (as suggested by D. Naylor) from effective (drained) material properties. As $K_f \to \infty$ incompressible behaviour of material found whether isotropic or anisotropic behaviour.
Quite large values of K_f can be used with isoparametric elements with reduced integration.
A useful sidestep into incompressible solution problems here exists.

III. Intermediate Time - Coupled Settlement

Solid phase equations are as before (case I) given in effective stress. Noting that p is not known this is included in analysis as unknown

$$p = \Sigma N_i p_i = \hat{\mathbf{N}} \mathbf{p}$$

$\mathbf{p}$ standing for appropriate nodal values (or parameters).

---

continued:

$$\mathbf{D}_f = \frac{1}{3} K_f \begin{bmatrix} 1 & 1 & 1 & 0 & 0 & 0 \\ & 1 & 1 & 0 & 0 & 0 \\ & & 1 & 0 & 0 & 0 \\ & & & 0 & 0 & 0 \\ & & & & 0 & 0 \\ & & & & & 0 \end{bmatrix}$$

$$\boldsymbol{\sigma} = \left[\mathbf{D}_e + \mathbf{D}_f \right] \boldsymbol{\varepsilon} .$$

Thus

$$\underset{\sim}{F}_h = - \quad \underset{\sim}{B}^T \begin{Bmatrix} 1 \\ 1 \\ 1 \\ 0 \\ 0 \\ 0 \end{Bmatrix} p \, d\Omega$$

$$= - \left(\int_\Omega \underset{\sim}{B}^T \begin{Bmatrix} 1 \\ 1 \\ 1 \\ 0 \\ 0 \\ 0 \end{Bmatrix} \hat{\underset{\sim}{N}} \, d\Omega \right) \underset{\sim}{p} = - \underset{\sim}{L} \, \underset{\sim}{p}$$

and in terms of effective stress equilibrium

$$\int_\Omega \underset{\sim}{B}^T \underset{\sim}{\sigma}_e \, d\Omega - \underset{\sim}{L} \, \underset{\sim}{p} + \underset{\sim}{F} = 0 \, ; \quad \underset{\sim}{\sigma}_e = \underset{\sim}{\sigma}_e(\underset{\sim}{\varepsilon}) \, ; \quad \underset{\sim}{\varepsilon} = \underset{\sim}{B} \, \underset{\sim}{u}$$

gives solid phase equation.

Fluid flow equation has now to be considered using Darcy's law

$$\frac{\partial}{\partial x} \left(K \frac{\partial p}{\partial x} \right) + \frac{\partial}{\partial y} \left(K \frac{\partial p}{\partial y} \right) + \frac{\partial \bar{\varepsilon}}{\partial t} - \frac{\eta}{K_f} \frac{\partial p}{\partial t} = 0$$

in which the last terms stated for changes in fluid stored.

Performing the standard discretisation we immediately find

$$\underset{\sim}{H} \, \underset{\sim}{p} - \int \hat{\underset{\sim}{N}}^T \frac{\partial \bar{\underset{\sim}{\varepsilon}}}{\partial t} \, d\Omega + \left(\int \hat{\underset{\sim}{N}}^T \frac{\eta}{K_f} \hat{\underset{\sim}{N}} \, d\Omega \right) \frac{\partial}{\partial t} \underset{\sim}{p} + \underset{\sim}{R} = 0$$

$\underset{\sim}{H}$ (see p. 298 ref. A) is dependent on k hence $\underset{\sim}{H} = \underset{\sim}{H} \left(\underset{\sim}{p} , \underset{\sim}{\varepsilon} \right)$

generally. As

$$\bar{\varepsilon} = \begin{Bmatrix} 1 \\ 1 \\ 1 \\ 0 \\ 0 \\ 0 \end{Bmatrix}^T \underset{\sim}{\varepsilon}$$

$$\int_\Omega \hat{N}^T \frac{\partial \bar{\varepsilon}}{\partial t} \, d\Omega = \left(\int_\Omega \hat{\underset{\sim}{N}}^T \begin{Bmatrix} 1 \\ 1 \\ 1 \\ 0 \\ 0 \\ 0 \end{Bmatrix}^T \underset{\sim}{B} \, d\Omega \right) \frac{\partial}{\partial t} \underset{\sim}{u} \equiv \underset{\sim}{L}^T \frac{\partial}{\partial t} \underset{\sim}{u}$$

with

$$S \equiv \int \hat{\underset{\sim}{N}}^T \frac{n}{K_f} \hat{\underset{\sim}{N}} \, d\Omega$$

we can write fluid equation as

$$\underset{\sim}{H} \underset{\sim}{p} - \underset{\sim}{L} \frac{\partial}{\partial t} \underset{\sim}{u} + \underset{\sim}{S} \frac{\partial}{\partial t} \underset{\sim}{p} + \underset{\sim}{R} = 0$$

(a) and (b) give a coupled system of equations which can be solved by time stepping.

For elastic behaviour $\underset{\sim}{\sigma}_e = \underset{\sim}{D}_e \underset{\sim}{\varepsilon}$

and

$$\int_\Omega \underset{\sim}{B}^T \underset{\sim}{\sigma}_e \, d\Omega = \underset{\sim}{K}_e \underset{\sim}{u}$$

and system can be rewritten as

$$\begin{bmatrix} \underset{\sim}{K}_e & -\underset{\sim}{L} \\ 0 & \underset{\sim}{H} \end{bmatrix} \begin{Bmatrix} \underset{\sim}{u} \\ \underset{\sim}{p} \end{Bmatrix} + \begin{bmatrix} 0 & 0 \\ -L^T & \underset{\sim}{S} \end{bmatrix} \frac{\partial}{\partial t} \begin{Bmatrix} \underset{\sim}{u} \\ \underset{\sim}{p} \end{Bmatrix} + \begin{Bmatrix} \underset{\sim}{F} \\ \underset{\sim}{R} \end{Bmatrix} = 0$$

which can be solved in time domain writing mid interval (Crank-

-Nicolson) scheme for instance

$$\begin{bmatrix} \mathbf{K}_e & -\mathbf{L} \\ -\mathbf{L}^T & \mathbf{A} \end{bmatrix} \begin{Bmatrix} \mathbf{u} \\ \mathbf{p} \end{Bmatrix}_{t+\Delta t} \begin{bmatrix} \mathbf{K}_e & -\mathbf{L} \\ \mathbf{L}^T & \mathbf{C} \end{bmatrix} \begin{Bmatrix} \mathbf{u} \\ \mathbf{p} \end{Bmatrix} + \begin{bmatrix} 2\mathbf{F}_{t+\frac{\Delta t}{2}} \\ \mathbf{B}_{t+\frac{\Delta t}{2}} \Delta t \end{bmatrix} = 0$$

$$\mathbf{A} = \mathbf{H}\,\Delta t/2 + \mathbf{S}$$

$$\mathbf{C} = \mathbf{H}\,\Delta t/2 + \mathbf{S}$$

This recurrence formula has been used by Sandhu and Wilson [24] and others later.

Note: As $\Delta t \rightarrow 0$ with $\mathbf{F}$ suddenly applied

$$2\mathbf{F}_{t+\frac{\Delta t}{2}} = \mathbf{F}$$

and with

$$\mathbf{u} = 0 \qquad \mathbf{p} = 0 \quad \text{at} \quad t = 0$$

we have

$$\begin{bmatrix} \mathbf{K}_e & -\mathbf{L} \\ -\mathbf{L}^T & \mathbf{A} \end{bmatrix} \begin{Bmatrix} \mathbf{u} \\ \mathbf{p} \end{Bmatrix} + \begin{Bmatrix} \mathbf{F} \\ \mathbf{0} \end{Bmatrix} = 0$$

for instant undrained conditions

$$\mathbf{A} \longrightarrow \mathbf{S}$$

($\mathbf{H}$ disappears as would be expected no drainage takes place) Apparently an alternative formulation of undrained problems is given. This from section II would be given as

$$\left(\int \mathbf{B}^T \left(\mathbf{D}_e + \mathbf{D}_F \right) \mathbf{B}\, d\Omega \right) \mathbf{u} + \mathbf{F} = 0$$

or

$$\left(\underset{\sim}{K}_e + \underset{\sim}{K}_h\right) \underset{\sim}{u} + \underset{\sim}{F} = 0$$

with

$$\underset{\sim}{K}_h = \int \underset{\sim}{B}^T \begin{Bmatrix} 1 \\ 1 \\ 1 \\ 0 \\ 0 \\ 0 \end{Bmatrix} \frac{K_f}{\eta} \begin{Bmatrix} 1 \\ 1 \\ 1 \\ 0 \\ 0 \\ 0 \end{Bmatrix}^T \underset{\sim}{B} \, d\Omega$$

eliminating $\underset{\sim}{p}$ from the combined equation system by multiplying second set by $\underset{\sim}{A}^{-1}$ a similar form will be obtained with a single equation for $\underset{\sim}{u}$. However the new system is valid for $K_f = \infty$, i.e. when fluid is totally incompressible. This formulation resembles incompressible formulation of Herrmann [25] but is more generally valid. Now for incompressible situation we have

$$\begin{bmatrix} \underset{\sim}{K}_e & -\underset{\sim}{L} \\ -\underset{\sim}{L} & 0 \end{bmatrix} \begin{Bmatrix} \underset{\sim}{u} \\ \underset{\sim}{p} \end{Bmatrix} + \begin{Bmatrix} \underset{\sim}{F} \\ 0 \end{Bmatrix} = 0$$

Non Linear Finite Element Equations

In many problems of structural or non-structural type non linear relations exist which result in discretised equations of form

$$\underset{\sim}{P}(\underset{\sim}{a}) + F = 0.$$

These present many difficulties in solution but physical intuition helps in deriving iterative, convergent schemes. Existence or uniqueness of solution can not be guaranteed.

Two Major Possibilities

(1) Direct use of Newton Raphson

Writing

$$d\underset{\sim}{P}(\underset{\sim}{a}) = \underset{\sim}{A}(\underset{\sim}{a})d\underset{\sim}{a}$$

where $\underset{\sim}{A}(\underset{\sim}{a})$ is a matrix

$$\begin{bmatrix} \frac{\partial P}{\partial a_1}, & \frac{\partial P}{\partial a_2}, & \dots \\ \frac{\partial P}{\partial a_1}, & \dots & \\ \dots & & \end{bmatrix}$$

we proceed to approximate starting from a guess $\underset{\sim}{a}^0$ and noting that

$$\underset{\sim}{P}(\underset{\sim}{a}^n + \Delta\underset{\sim}{a}^{n+1}) \approx P(a^n) + \underset{\sim}{A}(\underset{\sim}{a}^n)\Delta a^{n+1} = -\underset{\sim}{F}$$

we can iterate, and

$$\Delta \underset{\sim}{a}^{n+1} = - \underset{\sim}{A}(a^n)^{-1} \left(\underset{\sim}{F} + P(\underset{\sim}{a}^n)\right)$$

This in general is a convergent scheme. This use of N/R is expensive as $\underset{\sim}{A}$ has to be assembled and reinversed at each step.

Frequently convergence can be achieved either updating only after few iterations or keeping

$$\underset{\sim}{A}(a^n) = \underset{\sim}{A}(\underset{\sim}{a}^0)$$

throughout (modified N/R)

(2) <u>Step by step solution</u>

In many problems $\underset{\sim}{F}$ is a physical quantity (such as load) which has property

$$\underset{\sim}{a} = \underset{\sim}{a}^0 \quad \text{when} \quad F = \underset{\sim}{F}_0$$

and $\underset{\sim}{F}$ can be considered as a function of some parameter t

$$\underset{\sim}{F} = \underset{\sim}{F}_0 + \underset{\sim}{Q}\underset{\sim}{t}.$$

Differentiating non linear relation we have

$$\frac{d\underset{\sim}{P}}{dt} + \frac{d\underset{\sim}{F}}{dt} \equiv \underset{\sim}{A}(\underset{\sim}{a}) \frac{d\underset{\sim}{a}}{dt} + \underset{\sim}{Q} = 0$$

This is solved as a marching problem for known initial conditions $\left(a^0, F_0\right)$ (these initial conditions being often zero).

In solution of the marching problem various schemes can be adopted, e.g. Euler, Predictor-corrector, Hunga Kutta etc.

Simplest (Euler) often gives adequate result especially as existence of basic equation

$$\underset{\sim}{P}(\underset{\sim}{a}) + \underset{\sim}{F} = 0$$

allows error to be assessed at any stage and reinserted. This is probably most efficient scheme.

Further in some problems

$$t = \text{real time}$$

and

$$\underset{\sim}{P} \text{ is dependent not only on } \underset{\sim}{a} \text{ but on } \frac{\partial}{\partial t}\underset{\sim}{a}$$

so 'meaning' can be assigned.

For most efficient solution sometimes combination of N/R and marching process is used.

This is essential in cases where $\underset{\sim}{P}$ can only be determined in incremental sense (vide plasticity or creep, Nayak and Zienkiewicz). [26]

Example A Typical Non–Structural Problem

Arising from diffusion problem of non linear kind

$$\frac{\partial}{\partial x}\left(k\,\frac{\partial \Phi}{\partial x}\right)+\frac{\partial}{\partial y}\left(k\,\frac{\partial \Phi}{\partial y}\right)+Q=0$$

with

$$k = k(\Phi) \qquad Q = Q(\Phi)$$

discretised set with $\Phi = \underset{\sim}{N}\,\underset{\sim}{a}$

gives

$$\underset{\sim}{P}(\underset{\sim}{a}) + \underset{\sim}{F} = 0$$

$\underset{\sim}{F} = \underset{\sim}{F}_0$ prescribed for say boundary condition

and

$$\underset{\sim}{P}(\underset{\sim}{a}) \equiv \underset{\sim}{H}\,\underset{\sim}{a} + \int_\Omega \underset{\sim}{N}^T \underset{\sim}{Q}(\Phi)\,d\Omega$$

$$H_{ij} = \int_\Omega k(\Phi)\left(\frac{\partial N_i}{\partial x}\cdot\frac{\partial N_j}{\partial x}+\frac{\partial N_i}{\partial y}\cdot\frac{\partial N_j}{\partial y}\right)d\Omega = \int_\Omega k(\Phi)\,h'_{ij}\,d\Omega$$

on differentiation

$$\begin{Bmatrix} dP_1 \\ dP_2 \\ \vdots \end{Bmatrix} = \int_\Omega \left[\begin{matrix} kh'_{11}da_1 + kh'_{12}da_2 + \ldots + k'\left(\sum N_i da_i\right)a_1 + \\ kh'_{21}da_1 + kh'_{22}da_2 + \ldots + k'\left(\sum N_i da_i\right)a_1 + \\ \ldots\ldots \end{matrix}\right.$$

$$\left.\begin{matrix} +k'\left(\sum N_i d_i a_i\right)a_2 + \ldots + N_1 Q' \sum N_i da_i \\ +k'\left(\sum N_i d_i a_i\right)a_2 + \quad + N_2 Q' \sum N_i da_i \\ \ldots\ldots \end{matrix}\right]$$

with

$$\frac{dk}{d\Phi} = \Phi' \; ; \quad \frac{dQ}{d\Phi} = Q'$$

This yields the matrix $\underset{\sim}{A}$ with

$$A_{ij} = H_{ij} + \int_{\Omega} k' N_i \sum h_{ij} a_j \, d\Omega + \int_{\Omega} N_i Q' N_j \, d\Omega$$

first and last terms are symmetric but not the middle one.

N/R has been used with success in various problems. Variation of k occurs in seepage problems where Darcy's law does not hold. Saturating magnetic fields etc. Variation of Q in heat conduction, e.g. spontaneous ignition problem

$$Q = e^{\alpha \Phi}$$

Extremely accurate results obtained by Anderson [27] for points where solution ceases to exist (explosion), N/R works rapidly here.

Solid Mechanics

Small Strain – Inelastic Material – Discretised Equation

(Displacement formulation)

$$\underset{\sim}{P}(\underset{\sim}{a}) + \underset{\sim}{F} \equiv \int_{\Omega} \underset{\sim}{B}^T \underset{\sim}{\sigma}\, d\Omega + \underset{\sim}{F} = 0 = \underset{\sim}{\psi}$$

$\underset{\sim}{\sigma}$ a functional of displacement and time generally.

$\underset{\sim}{\psi}$ "residual", (unbalance), force

I. Unique Stress-strain Relation

$$\underset{\sim}{\sigma} = \underset{\sim}{\sigma}(\underset{\sim}{\varepsilon}) = \sigma(\underset{\sim}{u})$$

equilibrium equations define problems

both direct and incremental processes have been used

(1) Direct N/R

$$d\underset{\sim}{P} = d\int_{\Omega} \underset{\sim}{B}^T \underset{\sim}{\sigma}\, d\Omega = \int_{\Omega} \underset{\sim}{B}^T \frac{d\underset{\sim}{\sigma}}{d\underset{\sim}{\varepsilon}}\, d\underset{\sim}{\varepsilon}\, d\Omega$$

Noting that

$$\frac{d\underset{\sim}{\sigma}}{d\underset{\sim}{\varepsilon}} \equiv \underset{\sim}{D}_T \qquad \text{tangential stress-strain matrix}$$

$$d\underset{\sim}{\varepsilon} = \underset{\sim}{B}\, d\underset{\sim}{u}$$

$$d\underset{\sim}{P} = \left(\int \underset{\sim}{B}\ \ \underset{\sim}{D}_T\ B\, d\Omega\right) d\underset{\sim}{u} \equiv \underset{\sim}{K}_T\, d\underset{\sim}{u}$$

where K_T is tangential matrix

Starting from $\underset{\sim}{u}^0$ we iterate

$$\Delta \underset{\sim}{u}^{n+1} = -K_T^{n-1}\left(\underset{\sim}{F} + \int \underset{\sim}{B}_T \underset{\sim}{\sigma}^n d\Omega\right) = -\underset{\sim}{K}_T^{n-1} \cdot \underset{\sim}{\psi}^n$$

where $\underset{\sim}{\psi}$ is unbalanced force

Special process used in iteration $\underset{\sim}{K}_T = \underset{\sim}{K}_0$

K_0 - elastic matrix

as at every step $\underset{\sim}{\sigma}_e$ -elastic stress found exactly

$$\underset{\sim}{\psi} = -\int_\Omega \left(\underset{\sim}{\sigma} - \underset{\sim}{\sigma}_e\right) d\Omega + \underset{\sim}{F}$$

which is equivalent to introduction initial stresses (initial stress process, see ref. A, also ref. [28].

Process well convergent in practice for plasticity.

Process difficult to apply if locking material considered.

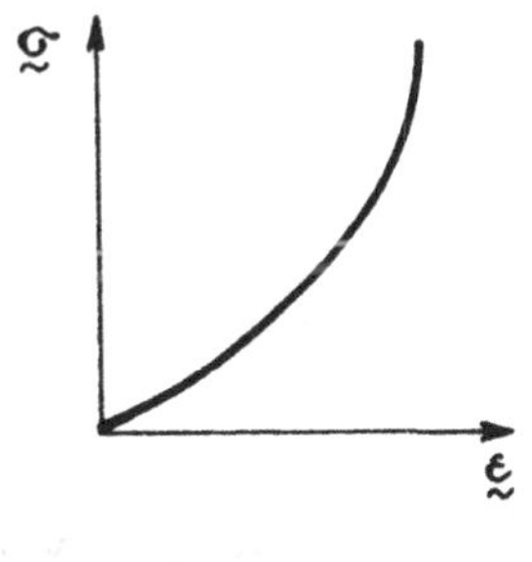

$$\frac{d\underset{\sim}{\sigma}}{d\underset{\sim}{\varepsilon}} \longrightarrow \infty$$

(2) Marching process

Involves written

$$\underset{\sim}{F} = \underset{\sim}{Q} \cdot t$$

$$\frac{d\underset{\sim}{P}}{dt} + \frac{d\underset{\sim}{F}}{dt} = \underset{\sim}{K}_T \frac{d}{dt} \underset{\sim}{u} + Q = 0$$

and incrementing t.

Thus, using Euler and starting from $\underset{\sim}{u} = \underset{\sim}{u}^0$

$$\Delta \underset{\sim}{u}^1 = \underset{\sim}{K}_T^{0^{-1}} \underset{\sim}{Q} \Delta t_1$$

$$\underset{\sim}{u}^1 = \underset{\sim}{u}^0 + \Delta \underset{\sim}{u}^1$$

To improve accuracy here

$$\underset{\sim}{\psi}^1 = \underset{\sim}{P}(u^1) + \underset{\sim}{F}^1 \qquad \text{is calculated}$$

and added to the 'load' $\underset{\sim}{Q} \Delta t_2$ in next step.

Thus

$$\Delta \underset{\sim}{u}^2 = \underset{\sim}{K}_T^{1^{-1}} (\underset{\sim}{Q} \Delta t_2 + \underset{\sim}{\psi}^1)$$

Note : Tangential Matrix = Marching Processes

Here we have

$$\underset{\sim}{K}_T(u) \frac{d}{dt} \underset{\sim}{u} + Q(t) = 0$$

or

$$\frac{d\underset{\sim}{u}}{dt} + \underset{\sim}{K}_T^{-1}(u) \, Q(t) = 0$$

Problem comes within the mathematical category of

$$\frac{dy}{dx} + f(x,y) = 0$$

and all techniques available for its solution apply.

1. Stability may present problems.

2. Looking at problem physically if at each step we evaluate stresses and displacements as if material were elastic then no residual should ever arise (excepting roundoff)
What is done in fact is that stresses are predicted independently and not by $\underset{\sim}{\sigma} = \underset{\sim}{D}_T \underset{\sim}{B} \underset{\sim}{u}$.

Stress Strain - Relationship Defined Incrementally Only

Now apparently direct method will not work as only $\underset{\sim}{D}_T$ is known at any time.

However $\underset{\sim}{\sigma} = \int d\underset{\sim}{\sigma}$ is integrated and total error ψ can still be found.

Plasticity falls into this category with $\underset{\sim}{D}_T$ available for any value of $\underset{\sim}{\sigma}$.

Incremental process used invariably but often combine large steps at Δt with N/R process to obtain residual $\underset{\sim}{\psi}$ and reduce it to zero by iteration.

Indeed success has been achieved using $\underset{\sim}{D}_T$ only for calculation of $\underset{\sim}{\sigma}$ and iterating with $\underset{\sim}{K}^0$ -elastic matrix only (initial stress process [28] , [26] .

Assumption implicit that error in integrating stress in t less than error in $\underset{\sim}{\psi}$.
Probably economic to use purely incremental process with minimum correction.

Initial Strain Processes

Law of behaviour such that only strain can be found from stress

$$\underset{\sim}{\varepsilon} = \underset{\sim}{\varepsilon}(\underset{\sim}{\sigma})$$

Here the iterative scheme has to be modified

$$\int_{\Omega} \underset{\sim}{B}^{T} \underset{\sim}{\sigma} \, d\Omega + F = 0$$

still valid but $\underset{\sim}{\sigma}$ can not be found explicitly.

If we write - as we always can

$$\underset{\sim}{\sigma} = \underset{\sim}{D}(\underset{\sim}{\varepsilon} - \underset{\sim}{\varepsilon}_0) \quad \text{or} \quad \begin{aligned} \underset{\sim}{\varepsilon}_0 &= \underset{\sim}{\varepsilon} - \underset{\sim}{D}\,\underset{\sim}{\sigma} \\ \underset{\sim}{\varepsilon}_0 &= \underset{\sim}{B}\,\underset{\sim}{u} - \underset{\sim}{D}\,\underset{\sim}{\sigma} \end{aligned}$$

original equation transforms to

$$\int_{\Omega} \underset{\sim}{B}^{T} \underset{\sim}{D} \, \underset{\sim}{\varepsilon} \, d\Omega - \int_{\Omega} \underset{\sim}{B}^{T} \underset{\sim}{D} \, \underset{\sim}{\varepsilon}_0 + \underset{\sim}{F} = 0$$

gives a simultaneous equation

$$\underset{\sim}{\psi} \equiv \underset{\sim}{K}\,\underset{\sim}{u} - \int \underset{\sim}{B}^{T} \underset{\sim}{D} \, \underset{\sim}{\varepsilon}_0 \, d\Omega + F = 0$$

$$\underset{\sim}{\sigma} = \underset{\sim}{D}(\underset{\sim}{\varepsilon} - \underset{\sim}{\varepsilon}_0) = \underset{\sim}{D}\,\underset{\sim}{B}\,\underset{\sim}{u} - \underset{\sim}{\varepsilon}_0 \,, \quad \underset{\sim}{\varepsilon} = \underset{\sim}{B}\,\underset{\sim}{u} = \underset{\sim}{\varepsilon}(\sigma).$$

This system must now be solved iteratively to determine $\underset{\sim}{u}$ and ε_0 . Taking

$$\underset{\sim}{\sigma}^{n+1} = \underset{\sim}{D}\left(\underset{\sim}{B}\,\underset{\sim}{u}^{n+1} - \varepsilon_0^{n}\right)$$

we find

$$\underset{\sim}{\varepsilon}^{n+1} = \underset{\sim}{\varepsilon}\left(\sigma^{n+1}\right) - \underset{\sim}{D}\,\underset{\sim}{B}\,\underset{\sim}{u}^{n+1}$$

taking $\underset{\sim}{\varepsilon}\left(\sigma^{n+1}\right)$ from constitutive law to be satisfied

ψ^{n+1} can now be found and

$$\Delta\,\underset{\sim}{u}^{n+1} = -\underset{\sim}{K}^{-1}\,\underset{\sim}{\psi}^{n+1} \qquad \text{determined}$$

This is the 'initial strain' process widely used.

(A) With $\underset{\sim}{D} = \underset{\sim}{D}_e$ corresponding to constant elastic matrix it is similar to the modified N/R process but not identical to it. With small deviations from elasticity process rapidly convergent.

It is interesting to note that the process is successful for solving problems with $\nu = 0{,}5$. Here

$$\underset{\sim}{\varepsilon} = \underset{\sim}{C}\,\underset{\sim}{\sigma}$$

where $\underset{\sim}{C}$ is defined but $\underset{\sim}{\bar{D}} = \underset{\sim}{C}^{-1}$ does not exist. Using as approximation $\underset{\sim}{D}$ solution can be found.

<u>Creep Problems</u> (Special case of initial strain)

Here constitutive law is of form

$$\underset{\sim}{\varepsilon} = \underset{\sim}{\bar{\varepsilon}}_c + \underset{\sim}{\bar{\varepsilon}}_e$$

where $\bar{\varepsilon}_e = \bar{\varepsilon}_e(\sigma)$ is of form which is invertible

$$\sigma = \sigma(\bar{\varepsilon}_e)$$

but

$$\varepsilon_c = \varepsilon_c(t, \sigma) \qquad t\text{ - time}.$$

Equilibrium demands

$$\int_\Omega B^T \sigma(\bar{\varepsilon})\, d\Omega + F(t) = 0$$

$$\bar{\varepsilon}_e = \varepsilon - \varepsilon_c = B u - \varepsilon_c(t, \sigma)$$

and this equation system need to be solved simultaneously. Differentiating with respect to parameter (real time), t, we have

$$\int B^T \frac{d\sigma}{d\bar{\varepsilon}_e} \frac{d\bar{\varepsilon}_e}{dt}\, d\Omega + \dot{F} = 0 \qquad \dot{F} = \frac{d}{dt} F$$

and

$$\frac{d\bar{\varepsilon}_e}{dt} = B \frac{du}{dt} - \dot{\varepsilon}_c(\sigma, t) \qquad \dot{\varepsilon}_c = \frac{d\varepsilon_c}{dt}$$

Noting that

$$\frac{d\sigma}{d\varepsilon_e} = D_T$$

we have written

$$K_T = \int_\Omega B^T D_T B\, d\Omega$$

$$\underset{\sim}{K}_T \frac{d}{dt} \underset{\sim}{u} - \int_\Omega \underset{\sim}{B}^T \underset{\sim}{D}_T \dot{\underset{\sim}{\varepsilon}}_c \, d\Omega + \dot{\underset{\sim}{F}} = 0$$

$$\frac{d}{dt} \underset{\sim}{\sigma} = D_T \frac{d}{dt} \bar{\underset{\sim}{\varepsilon}} = \underset{\sim}{D}_T \left(\underset{\sim}{B} \frac{d}{dt} \underset{\sim}{u} - \dot{\varepsilon}_c \right)$$

$$D_T = D_T(\sigma) \; ; \quad \dot{\varepsilon}_c = \dot{\varepsilon}_c(\sigma, t).$$

This system is similar to the one encountered in initial strain process. Indeed $\underset{\sim}{\varepsilon}_c$ (or $\Delta \underset{\sim}{\varepsilon}_c$) plays directly part of initial strain.

1) Now however explicit forms ot the creep strain rate are given

$$\dot{\varepsilon}_c = \dot{\varepsilon}(\sigma, t)$$

2) Solution most conveniently found by taking small time steps and using Euler, predictor corrector, Runge Kutta or other methods.

Single Step Forward (Euler) Procedure

starting from known value of $\underset{\sim}{u}_0$ and $\underset{\sim}{\sigma}_0$ at $t = t_0$

(1) find $$\dot{\varepsilon}_c^0 = \dot{\varepsilon}_c^0 (\sigma_0, t_0)$$

(2) find $$\Delta \underset{\sim}{u} = \underset{\sim}{K}_T^{0-1} \left[\int_\Omega B^T D_T \dot{\varepsilon}_c^0 \Delta t \, d\Omega - \Delta \underset{\sim}{F} \right]$$

(3) find $$\Delta \underset{\sim}{\sigma} = \underset{\sim}{D}_T \left(\underset{\sim}{B}\, \Delta \underset{\sim}{u} - \dot{\underset{\sim}{\varepsilon}}{}_c^{o} \Delta t \right)$$

(4) find $$\underset{\sim}{u}^{\Delta t} = \underset{\sim}{u}^{o} + \Delta \underset{\sim}{u} \quad , \quad \underset{\sim}{\sigma}^{\Delta t} = \underset{\sim}{\sigma}^{o} + \Delta \underset{\sim}{\sigma}$$

Correction I -

a second iteration within the time step can now be introduced, take

$$\dot{\varepsilon}_c^{*} = \dot{\varepsilon}_c \left(t \, , \sigma_o + \frac{\Delta \sigma}{2} \right)$$

and recompute

$$\Delta \underset{\sim}{u}^{*} \text{ and } \Delta \sigma^{*}$$

Correction II -

recompute $\dot{\varepsilon}_c^{*}$

then find $\Delta \underset{\sim}{\sigma}^{**} = D_T \left(B \, \Delta u - \dot{\varepsilon}_o^{*} \Delta t \right)$; $\underset{\sim}{\sigma}^{\Delta t} = \underset{\sim}{\sigma}^{o} + \Delta \underset{\sim}{\sigma}^{**}$

but do not resolve equation taking equilibrium error

$$\underset{\sim}{\psi}_1 = \int_{\Omega} \underset{\sim}{B}^T \underset{\sim}{\sigma}^{\Delta t} d\Omega - \underset{\sim}{F}$$

and adding this error to next time step.

This is quite economical as operation is combined and done together with evaluation of $\int_{\Omega} \underset{\sim}{B}^T \underset{\sim}{D}_T \dot{\underset{\sim}{\varepsilon}}_c \Delta t \, d\Omega$ for next time step.

Some Typical Details of a Few Constitutive Relations

These are very brief and deal with

Non-associate (general) plasticity

Visco-plasticity

Limited cracking behaviour

These three classes are universally useful and comprise a general 'library'.

Plasticity Stress Strain Relations

Summary of constitutive relations only is given here in terms needed for FEM (see also ref. [26] or ref. A).

1. Yield surface

$$F(\underset{\sim}{\sigma}, \kappa)$$

defines limits of elastic behaviour (k hardening parameter)

$$F < 0 \quad \text{elastic}$$

$$F = 0 \quad \text{plastic}$$

2. Direction of straining

Plastic strain assumed to proceed in direction normal to plastic $Q(\underset{\sim}{\sigma})$ potential i.e.

$$\underset{\sim}{\varepsilon}_p = \lambda \frac{\partial Q}{\partial \underset{\sim}{\sigma}} = \lambda \underset{\sim}{a}^*$$

$$\underset{\sim}{a}^* = \begin{Bmatrix} \dfrac{\partial Q}{\partial \sigma_1} \\ \dfrac{\partial Q}{\partial \sigma_2} \\ \vdots \end{Bmatrix}$$

Special case $Q = F$
associated rule

$$\underset{\sim}{a}^* = \underset{\sim}{a}$$

λ - arbitrary positive constant

3. Total strain - elastic strain and plastic strain

$$d\varepsilon = d\underset{\sim}{\varepsilon}_e + d\underset{\sim}{\varepsilon}_p$$

$$d\underset{\sim}{\varepsilon}^e = D^{-1} d\sigma.$$

From above we can determine

$$\underset{\sim}{D}_T \; ; \quad d\underset{\sim}{\varepsilon}^p \; ; \quad \lambda$$

During plastic deformation

$$F = 0$$

i.e.

$$\frac{\partial F^T}{\partial \underset{\sim}{\sigma}} d\underset{\sim}{\sigma} + \frac{\partial F}{\partial \varkappa} d\varkappa = 0$$

inserting

$$\frac{\partial F}{\partial \underset{\sim}{\sigma}} = \underset{\sim}{a} \qquad A = -\frac{1}{\lambda} \frac{\partial F}{\partial \varkappa} d\varkappa$$

Thus summarising

$$\underset{\sim}{a}^T d\underset{\sim}{\sigma} - A\lambda = 0 \qquad (1)$$

$$d\underset{\sim}{\varepsilon}_p = \lambda \, \underset{\sim}{a}^* \qquad (2)$$

$$d\underset{\sim}{\varepsilon}_e = \underset{\sim}{D}^{-1} \, d\underset{\sim}{\sigma} \qquad (3)$$

$$d\underset{\sim}{\varepsilon} = d\underset{\sim}{\varepsilon}_e + d\underset{\sim}{\varepsilon}_p \qquad (4)$$

Useful relations

From (1) $\times$ a^* and (2)

$$\frac{a^* a^T}{A} \, d\underset{\sim}{\sigma} = d\underset{\sim}{\varepsilon}^p \qquad \text{Basic relation}$$

From (4) $\times$ $\underset{\sim}{a}^T \underset{\sim}{D}$ and (3), (2)

$$\underset{\sim}{a}^T \underset{\sim}{D} \, d\underset{\sim}{\varepsilon} = \underset{\sim}{a}^T d\underset{\sim}{\sigma} + \underset{\sim}{a}^T \underset{\sim}{D} \, \underset{\sim}{a}^* \lambda$$

using (1)

$$\underset{\sim}{a}^T \underset{\sim}{D} \, d\underset{\sim}{\varepsilon} = \lambda A + \lambda \, \underset{\sim}{a}^T \underset{\sim}{D} \, \underset{\sim}{a}^*$$

Hence

$$\lambda = \underset{\sim}{a}^T \underset{\sim}{D} \, d\varepsilon / C \quad , \qquad C = A + a^T D \, a^*$$

As from (4), (3) and (2)

$$d\underset{\sim}{\varepsilon} = \underset{\sim}{D}^{-1} d\underset{\sim}{\sigma} + \underset{\sim}{a}^* \underset{\sim}{\lambda} = \underset{\sim}{D}^{-1} d\underset{\sim}{\sigma} + \frac{\underset{\sim}{a}^* \underset{\sim}{a}^T \underset{\sim}{D}}{C} \, d\underset{\sim}{\varepsilon}$$

$$d\underset{\sim}{\sigma} = \left(\underset{\sim}{D} + \underset{\sim}{D} \, \underset{\sim}{a}^* \underset{\sim}{a}^T D / C \right) d\underset{\sim}{\varepsilon} \, .$$

Hence

$$\underset{\sim}{D}_T = \underset{\sim}{D} + \underset{\sim}{D}^p \ , \qquad \underset{\sim}{D}^p = \underset{\sim}{D}\,\underset{\sim}{a}^*\,\underset{\sim}{a}^T D \,/\, A + \underset{\sim}{a}^T \underset{\sim}{D}\,\underset{\sim}{a} .$$

In preceding $\varkappa$ was the usual isotropic hardening parameter. For ideal plasticity $A = 0$.

Visco Plasticity

Here we shall take a law defining the plastic/ /creep strain rate as a function of the distance from the yield surface. Thus if

$$\text{if} \quad \begin{array}{ll} F < 0 & \dot{\underset{\sim}{\varepsilon}}_{vp} = 0 \\ F \geq 0 & \dot{\underset{\sim}{\varepsilon}}_{vp} \neq 0 . \end{array}$$

Further, direction of visco-plastic strain will be taken in a direction specified by plastic potential $Q(\underset{\sim}{\sigma})$.

Thus

$$\dot{\underset{\sim}{\varepsilon}}_{vp} = \frac{\partial Q}{\partial \underset{\sim}{\sigma}} \Phi (< F >)$$

when $< F >$ means that for negative values of $F, \Phi = 0$. Typical Φ used are

$$\Phi = \mu < F > , \quad \Phi = \mu < F >^n , \quad \Phi = \mu\, a^{<F>}$$

if

$$F = Q$$

we have associated visco-plasticity.

As in plasticity a 'strain hardening' parameter $\varkappa$ can be introduced.

1) Visco-plasticity program includes creep behaviour of typical metals (if very small yield stress used) and
2) includes an alternative for solution of plasticity problems
3) In solution technique we use a simple initial strain formulation as that defined in creep problems.

Note: Comparison of modified N/R process (initial stress) in plasticity and solution of a visco-plastic problem.

In 'initial stress'(or modified N/R process) of elastic plastic solution we solve with original k matrix at each step for a residual force i.e.

$$\Delta \underset{\sim}{u}^{n+1} = - \underset{\sim}{K}_0^{-1} \left(\int_\Omega \underset{\sim}{B}^T (\underset{\sim}{\sigma}_n - \underset{\sim}{\sigma}_n^p) \, d\Omega \right)$$

where σ_n was total stess reached at start of iteration and $\underset{\sim}{\sigma}_n^p$ corresponding plastic stress increment.

In a typical step of incremental, initial strain analysis of creep

$$\Delta \underset{\sim}{u}^{n+1} = - K_0^{-1} \int_\Omega B^T D \, \dot{\underset{\sim}{\varepsilon}}_{vp} \Delta t \, d\Omega$$

if $\dot{\underset{\sim}{\varepsilon}}_{vp} = \mu < F > a^*$ (linear viscous law)

both steps identical if

$$\underset{\sim}{\sigma}_n - \underset{\sim}{\sigma}_n^p = \underset{\sim}{D} \, \mu < F > \underset{\sim}{a}^* \Delta t$$

in uniaxial problem $F = \underset{\sim}{\sigma} - \underset{\sim}{\sigma}^p$ exactly but generally steps will not be identical though similar for a suitable choice of Δt.

Material With Limited Tensile Resistance

Rock/concrete etc.

Here constitutive law is elastic until max. principal stress. After that point that particular principal stress put = 0 and $\underset{\sim}{D}_T$ such that changes of stress in that direction = 0. (further details in ref. A).

Transient - Step by Step Solution

(Linear and non-linear situations. Possible use to find steady state solution. Dynamic relaxation).

Parabolic Problems

Consider

$$\frac{\partial}{\partial x}\left(k\,\frac{\partial \Phi}{\partial x}\right) + \frac{\partial}{\partial y}\left(k\,\frac{\partial \Phi}{\partial y}\right) + Q + c\,\frac{\partial \Phi}{\partial t} = 0$$

with Φ known at $t = 0$ and suitable spatial boundary conditions

(K, Q and c depend on Φ and t)

Standard (Galerkin) discretization with $\Phi = \sum N_i a_i$

gives

$$\underset{\sim}{H}\,\underset{\sim}{a} + \underset{\sim}{C}\,\dot{\underset{\sim}{a}} + \underset{\sim}{F} = 0$$

with

$$H_{ij} = \int_{\Omega} k\left(\frac{\partial N_i}{\partial x}\cdot\frac{\partial N_j}{\partial x} + \frac{\partial N_i}{\partial y}\cdot\frac{\partial N_j}{\partial y}\right) d\Omega$$

$$C_{ij} = \int_{\Omega} N_i\, c\, N_j\, d\Omega$$

$$F_i = \int_{\Omega} N_i\, Q\, d\Omega$$

Various numerical schemes can be used

(a) Simple finite difference (Euler, Crank Nicholson etc.)

(b) Various Runge-Kutta formula

(c) Various finite elements in time

We shall consider only (a).

(a) Mid-interval (Crank-Nicholson) scheme

Write

$$\underset{\sim}{a} = \left(\underset{\sim}{a}_t + a_{t+\Delta t}\right)/2$$

$$\dot{\underset{\sim}{a}} = \left(a_{t+\Delta t} - a_t\right)/\Delta t$$

$$\left(\underset{\sim}{H}_m \Delta t + 2\underset{\sim}{C}_m\right)\underset{\sim}{a}_{t+\Delta t} = -2\Delta t \underset{\sim}{F}_m - \left(\underset{\sim}{H}_m \Delta t - 2\underset{\sim}{C}_m\right)\underset{\sim}{a}_t = \underset{\sim}{\psi}_t$$

$$\underset{\sim}{a}_{b+\Delta b} = \left(H_m \Delta t + 2C_m\right)^{-1}\underset{\sim}{\psi}_e .$$

In above H_m, C_m and F_m stand for mid interval values. In linear problem H_m and C_m are constant and $\underset{\sim}{F} = \underset{\sim}{F}(t)$ can be determined. In non linear problem we can either take $\underset{\sim}{H}_m = \underset{\sim}{H}_t$ etc. or iterate.

For linear problems

$$\left(\underset{\sim}{H}_m \Delta t + 2\underset{\sim}{C}_m\right)$$

can be inverted once but with non-linear situations successive steps very costly.

This scheme (at least for linear problems) unconditionally stable.

(b) Euler-forward Difference Scheme

Write
$$\underset{\sim}{a} = \underset{\sim}{a}_t$$

$$\dot{\underset{\sim}{a}} = \left(\underset{\sim}{a}_{t+\Delta t} - \underset{\sim}{a}_t\right)/\Delta t$$

$$\underset{\sim}{C}_t \underset{\sim}{a}_{t+\Delta t} = \underset{\sim}{C}_t \underset{\sim}{a}_t - \Delta t \underset{\sim}{F}_t - \underset{\sim}{H}_t \underset{\sim}{a}_t \Delta t = \underset{\sim}{\psi}_t$$

$$\underset{\sim}{a}_{t+\Delta t} = \underset{\sim}{C}_t^{-1} \underset{\sim}{\psi}_t .$$

Error in this scheme is larger but form a little simpler. Non-linearity does not require iteration. But as we still have to invert $\underset{\sim}{C}_t$ which is of same band width as $\underset{\sim}{H}$ the computation cost is the same. Also instability will now occur if Δt too large.

Note: The nomenclature of finite differences where (a) is called implicit while (b) explicit can not be here applied truly as both are implicit and the popularity of (b) procedure appears not to be present.

Note on Stability Consideration

If $\underset{\sim}{F}_t = 0$ the solution for successive steps of t should be a decaying one.

Writing

$$\underset{\sim}{a}_{t+\Delta t} = \left(\underset{\sim}{I} - \underset{\sim}{C}_t^{-1} \underset{\sim}{H}_t \Delta t\right) \underset{\sim}{a}_t$$

for Euler scheme or for mid interval scheme

$$\underset{\sim}{a}_{t+\Delta t} = -\left(\underset{\sim}{I} - \left(\underset{\sim}{H}_m \Delta t + 2 \underset{\sim}{C}_m\right)^{-1} 4 \underset{\sim}{C}_m\right) a_t$$

(adding $4C_m - 4C_m$ on RHS)

we note the general form

$$\underset{\sim}{a}_{t+\Delta t} = (\underset{\sim}{I} - \underset{\sim}{M})\underset{\sim}{a}_t .$$

If we want solution to decay, we should have

$$a_{t+2\Delta} - a_{t+\Delta} = \lambda (a_{t+\Delta} - a_t)$$

with $\lambda < 1$.

Above can be written as

$$\underset{\sim}{M} \cdot \underset{\sim}{M}\, \underset{\sim}{a}_t = \lambda\, \underset{\sim}{M}\, \underset{\sim}{a}_t$$

or

$$\underset{\sim}{M}\, (\lambda \underset{\sim}{I} - M)\, \underset{\sim}{a}_t = 0$$

λ is seen to be the eigenvalue of

$$\underset{\sim}{M} - \lambda \underset{\sim}{I} = 0$$

and stability condition is that the largest eigenvalue be less than 1. It can be shown that this is always so for Crank-Nicholson scheme but is dependent on Δt and element subdivision in Euler forward difference. (For estimates see Irons-Wright Patterson 1971).

Diagonalization of $\underset{\sim}{C}$ Matrix

The $\underset{\sim}{C}$ matrix corresponds to the 'retention' property of heat (or any other retentive physical quantity). If physically this were such as to be considered concentrated at nodes, the $\underset{\sim}{C}$ matrix would be diagonal and then the Euler difference scheme would become explicit. (Just as it always is in finite difference models).

For such a case several advantages acrue to the second scheme.

(a) The 'inverse', $\underset{\sim}{C}^{-1}$, is trivial (time advantage in computation).

(b) The inverse $\underset{\sim}{C}^{-1}$ and operation of finding $a_{i(t+\Delta t)}$ can be accomplished without assembly of any equations (storage advantage).

(c) Any non linearities present in $\underset{\sim}{C}$ or $\underset{\sim}{H}$ can be dealt with immediately and do not increase time of a single step calculation.

The last advantage is very substantial over the mid-difference process where for non-linear problems a matrix has to be continuously updated, and resolved.

The disadvantage obviously is the possibility of instability, necessitating use of small time steps. Simplicity of programming for 'lumped' $\underset{\sim}{C}$ operation are such a serious advantage as to make the process increasingly popular.

Steady Wave - non linear solution (Dynamic - relaxation)

The process (providing Δt sufficiently small will converge to the non-linear steady state solution. Here it represents a possible way of obtaining steady state solutions for non-

linear problems similar to that used in a finite difference process known as Dynamic-Relaxation (Otter J. R. M. Nucl. Struc. Eng. v.1. p. 61-75, 1965)
see also (R. D. Lynch, S. Kelsey and C. Saxe: "The application of dynamic relaxation to the finite element method of structural analysis" Univ. of Notre Dame, Tech. Report 1968).

More on "Lumping" Processes

In simple elements the method of 'lumping' is physically obvious and indeed advantages of using 'consistent' matrices are not great. In complex elements it has been repeatedly shown in vibration problems that the error in lumping increases.

It is logical to arrive at consistent matrices in a computer program using standard processes and to diagonalize later.
One such process is to add all terms and place on diagonal

$$\bar{C}^e_{ii} = \sum C^e_{ij} \qquad \bar{C}_{ij} = 0 \qquad i \neq j .$$

This gives correct 'response' for uniform changes of Φ but may lead to non-positive matrices in complex elements.
Example of a (linear) one dimensional problem.

Consider

$$K\frac{\partial^2\Phi}{\partial x^2}+C\frac{\partial\Phi}{\partial t}+Q(x,t)=0\,;\quad \Phi=0\quad x=0\,,\,nL.$$

Domain $0<x<nL$ divided into elements of length L with linear interpolation functions

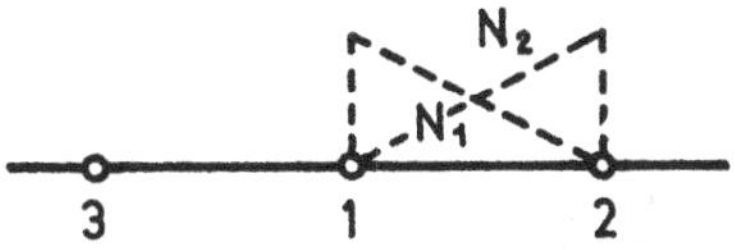

$$h_{ij}^e=\int_0^L\frac{\partial N_i}{\partial x}\frac{\partial N_j}{\partial x}dx$$

$$h_{11}^e=K/L$$

$$\frac{\partial N_1}{\partial x}=-1/L\qquad h_{12}^e=h_{21}^e=-K/L$$

$$N_1=(L-x)/L\qquad N_2=x/L\qquad \frac{\partial N_2}{\partial x}=1/L\qquad h_{22}^e=K/L$$

If Q const in element

$$C_{11}^e=Lc/3$$

$$F_1=F_2=QL/2\qquad C_{12}^e=C_{21}^e=Lc/6$$

$$F_i=\int_0^L N_i\,Q\,dx\qquad C_{ij}=\int_0^L N_i\,c\,N_j\,dx\qquad C_{22}^e=Lc/3$$

On assembly (assuming all elements identical) a typical equation is

$$2k/L\,\Phi_1-(\Phi_2+\Phi_3)k/L+\frac{2}{3}Lc\,\dot\Phi_1+(\dot\Phi_2+\dot\Phi_3)Lc/6+QL=0.$$

Using a forward marching scheme we have

$$\frac{2}{3}Lc\,\Phi_1+\frac{1}{6}Lc(\Phi_2+\Phi_3)\Big|_{t+\Delta t}=\frac{2}{3}Lc\,\Phi_1+\frac{1}{6}Lc(\Phi_2+\Phi_3)-$$

$$-QL\,\Delta t-\frac{2k}{L}\Phi_1+\frac{k}{L}(\Phi_2+\Phi_3)\Big|_t.$$

This requires solution of a tridiagonal equation system for $\Phi_{t+\Delta}$. Should we lump by adding elements of $\underset{\sim}{C}$ matrix we get

$$Lc\,\Phi_1\Big|_{t+\Delta} = Lc\,\Phi_1^t - QL\Delta t - \frac{k}{L}\Delta t\left(2\Phi_1 - \Phi_2 - \Phi_3\right)\Big|_t$$

which is identical with a forward marching finite difference scheme (known to be stable for)

$$Lc = \frac{2k\Delta t}{L} \quad \text{or} \quad \Delta t = \frac{L^2 c}{2k}$$

vide S. Crandal, Engineering Analysis McGraw-Hill 1965. Lumping is certainly useful and correct in this example.

Other Artifices for Reduction of Work in Time Stepping

The smallness of time interval involved in 'explicit' schemes and the need for reinversion of the matrix in more conventional mid difference scheme can be improved by reduction of degrees of freedom in original equations.

The procedure is similar to that used in 'eigenvalue economising' (see ref. A p. 350).

Let the parameters $a_i = a_i(t)$ be expressed (constrained) in terms of a smaller number $b_i = b_i(t)$ with a transformation

$$\underset{\sim}{a} = \underset{\sim}{T}\,\underset{\sim}{b}$$

Then equations

$$\underset{\sim}{H}\,\underset{\sim}{a} + \underset{\sim}{C}\,\underset{\sim}{\dot{a}} + \underset{\sim}{F} = 0$$

will be transformed to

$$\underset{\sim}{\hat{H}}\,\underset{\sim}{b} + \underset{\sim}{\hat{C}}\,\underset{\sim}{\dot{b}} + \underset{\sim}{\hat{F}} = 0$$

with

$$\underset{\sim}{\hat{H}} = \underset{\sim}{T}^T \underset{\sim}{H}\,\underset{\sim}{T} \qquad \underset{\sim}{\hat{C}} = \underset{\sim}{T}^T \underset{\sim}{C}\,\underset{\sim}{T} \qquad \underset{\sim}{\hat{F}} = \underset{\sim}{T}^T \underset{\sim}{F}$$

must be capable of approximating the distribution reasonably. It is possible to do this by obtaining suitable steady state solutions for prescribed values of nodal 'displacement' **b** .

The reduced problem is clearly economical as

(I) matrix to be inverted and number of unknowns to be handled if one time step is reduced.

(II) as high eigenvalues are eliminated larger time step can be used without instability (however - as lumping can now not be recommended - mid interval ascheme which is unconditionally stable will be used).

A special Form of a Parabolic Equation

An interesting equation arises in problems of convective diffusion.

Here the equation is of the type

$$\frac{\partial \Phi}{\partial t} + u\frac{\partial \Phi}{\partial x} + v\frac{\partial \Phi}{\partial y} + \frac{\partial}{\partial x}\left(k_x \frac{\partial \Phi}{\partial x}\right) + \frac{\partial}{\partial y}\left(k_y \frac{\partial \Phi}{\partial y}\right) = 0$$

in which u and v are velocities - which are prescribed functions os position and k_x, k_y are turbulent diffusion coefficients. Treating discretisation in the usual Galerkin manner we arrive at a similar situation to that previously described, i.e. with $\Phi = \sum N_i a_i$

$$(\underset{\sim}{H} + \underset{\sim}{D})\underset{\sim}{a} + \underset{\sim}{C}\dot{\underset{\sim}{a}} + \underset{\sim}{F} = 0.$$

The new feature is the presence of the matrix $\underset{\sim}{D}$ which is not symmetrical.

$$D_{ij} = \int_\Omega \left(N_i u \frac{\partial N_j}{\partial x} + N_i v \frac{\partial N_j}{\partial y}\right) d\Omega.$$

Treating $\underset{\sim}{H} + \underset{\sim}{D} = \hat{\underset{\sim}{H}}$ the situation becomes identical to that previously described. But in mid-interval formulation now an unsymmetric matrix need to be inverted. This present computational difficulties and the Euler, forward differences, scheme again is rather appealing.

It is easy to see that this dissymmetry is inherent if a finite difference approximation to the equation is made in a one-dimensional case

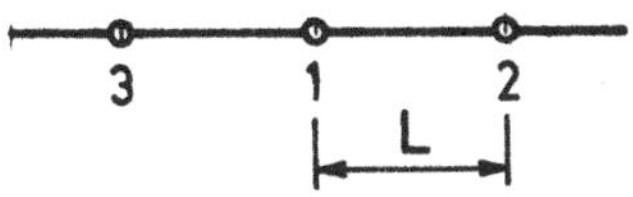

$$\frac{\partial \Phi}{\partial t} + u \frac{\partial \Phi}{\partial x} + \frac{\partial}{\partial x}\left(k \frac{\partial \Phi}{\partial x}\right) = 0$$

$$\dot{\Phi}_1 + \frac{u(\Phi_2 - \Phi_3)}{2L} + \frac{k}{L^2}(\Phi_2 + \Phi_3 - 2\Phi_1) = 0$$

In an explicit scheme

$$\Phi_1 \Big|_{t+\Delta t} = \Phi_1 - \frac{u}{2L}(\Phi_2 - \Phi_3)\Delta t + \frac{k \Delta t}{L^2}(\Phi_2 + \Phi_3 - 2\Phi_1)\Big|_t .$$

Thus explicit scheme proceeds on same basis as before. A possibility of obtaining symmetric equations (at expense of requiring C_1 continuity) exists if we use mean square error minimisation.

Hyperbolic Equations. Non - linear and Linear

Transient, Non - linear, Dynamic Problems

Basic equation here

$$\int_{\Omega} \underset{\sim}{B}^T \underset{\sim}{\sigma}\, d\Omega + \underset{\sim}{C}\dot{\underset{\sim}{u}} + \underset{\sim}{M}\ddot{\underset{\sim}{u}} + \underset{\sim}{F} = 0$$

$$\underset{\sim}{\sigma} = \underset{\sim}{\sigma}(\varepsilon) = \underset{\sim}{\sigma}(\underset{\sim}{u}) \qquad \int_{\Omega} \underset{\sim}{B}^T \underset{\sim}{\sigma}\, d\Omega = \underset{\sim}{R}(\underset{\sim}{u})$$

where $\underset{\sim}{M}$ is the mass matrix (always linear) and $\underset{\sim}{C}$ is damping matrix which could well be non linear is $\underset{\sim}{C} = \underset{\sim}{C}(\dot{\underset{\sim}{u}})$. Problems of above kind being second order will have both $\underset{\sim}{u}$ and $\dot{\underset{\sim}{u}}$ specified at $t = 0$ (initial condition).

For linear problems we can write a special form

$$\underset{\sim}{K}\underset{\sim}{u} + \underset{\sim}{C}\dot{\underset{\sim}{u}} + \underset{\sim}{M}\ddot{\underset{\sim}{u}} + F = 0.$$

Again various numerical schemes can be developed from

(a) finite difference expression

(b) time integration

(c) finite element in time

A brief survey of (a) type processes is given here. For more details see R.H. Clough, NATO Adv. Study Inst. Lisbon 1971.

For convenience we adopt notation

$$\underset{\sim}{u}_t = \underset{\sim}{u}_0 \qquad \underset{\sim}{u}_{t+\Delta t} = \underset{\sim}{u}_1 \qquad \underset{\sim}{u}_{t-\Delta t} = \underset{\sim}{u}_{-1} \qquad \text{etc.}$$

Time stepping procedures assume that at time t we know

$$\underset{\sim}{u}_0 \quad \text{and} \quad \dot{\underset{\sim}{u}}_0$$

or alternatively

$$\underset{\sim}{u}_0 \quad \text{and} \quad \underset{\sim}{u}_{-1}$$

For some schemes (after the first step) we know

$$\underset{\sim}{u}_0 \, , \; \underset{\sim}{u}_{-1} \, , \; \underset{\sim}{u}_{-2} \; \text{etc.}$$

and better extrapolation is possible.

We note also that in non-linear problems:

1. It is possible always to evaluate non-linear form

$$\int_\Omega \underset{\sim}{B}^T \underset{\sim}{\sigma} \, d\Omega \Big|_t = R_t$$ if $\underset{\sim}{u}_t$ and all previous values are known.

2. That sometimes it may be useful to write

$$\int_\Omega \underset{\sim}{B}^T \underset{\sim}{\sigma} \, d\Omega \Big|_{t+\Delta t} = \int_\Omega \underset{\sim}{B}^T \underset{\sim}{\sigma} \, d\Omega \Big|_t + \underset{\sim}{K}_T \left(\underset{\sim}{u}_{t+\Delta t} - \underset{\sim}{u}_t \right)$$

$$= \underset{\sim}{R}_t + \underset{\sim}{K}_t \, \Delta \underset{\sim}{u}_{t+\Delta t} \, .$$

Equation becomes

$$\underset{\sim}{M} \, \ddot{\underset{\sim}{u}} + \underset{\sim}{C} \, \dot{\underset{\sim}{u}} + \underset{\sim}{R}(u) + F = 0$$

I. Finite Difference Schemes

(a) Central difference

$$\ddot{\underset{\sim}{u}}_0 = -\underset{\sim}{M}^{-1}\left(\underset{\sim}{C}\,\dot{\underset{\sim}{u}}_0 + \underset{\sim}{R}_0 + \underset{\sim}{F}_0\right)$$

can be calculated as

$$\ddot{\underset{\sim}{u}}_0 = \left(\underset{\sim}{u}_1 - 2\,\underset{\sim}{u}_0 - \underset{\sim}{u}_{-1}\right)/\Delta t^2.$$

Find

$$\underset{\sim}{u}_1 = 2\,\underset{\sim}{u}_0 - \underset{\sim}{u}_{-1} - \Delta t^2 M^{-1}\left(\underset{\sim}{C}\,\dot{\underset{\sim}{u}} + \underset{\sim}{R}_0 + \underset{\sim}{F}_0\right).$$

Find $\dot{u}_1$ usign backward difference

$$\dot{\underset{\sim}{u}}_1 = \left(\underset{\sim}{u}_1 - \underset{\sim}{u}_0\right)/\Delta t.$$

Self starting, simple scheme unstable if Δt larger than T_{min}/n (T_{min} - min. natural period).

Used with success in many finite difference schemes.

No artificial damping. Ideal for impact type problems with high non-linearity.

Lumped $\underset{\sim}{M}$ (diagonal) simplifies calculation and limits storage. (vide Large elasto plastic shell transients) [29] . L. Marino, J. W. Leech and E. A. Witmer, J. Applied Mechanics, June 1971)

(b) Backward - higher order methods (J. Houbolt)

Expression for $\ddot{u}_1$ obtained by linear extrapolation of $\ddot{u}_0$ and $\ddot{u}_{-1}$ i.e.

$$\ddot{\underset{\sim}{u}}_1 = 2\ddot{\underset{\sim}{u}}_0 - \ddot{\underset{\sim}{u}}_{-1} = $$
$$= \left[2(u_1 - 2u_0 + u_{-1}) - (u_0 - 2u_{-1} + u_{-2})\right] / \Delta t^2$$

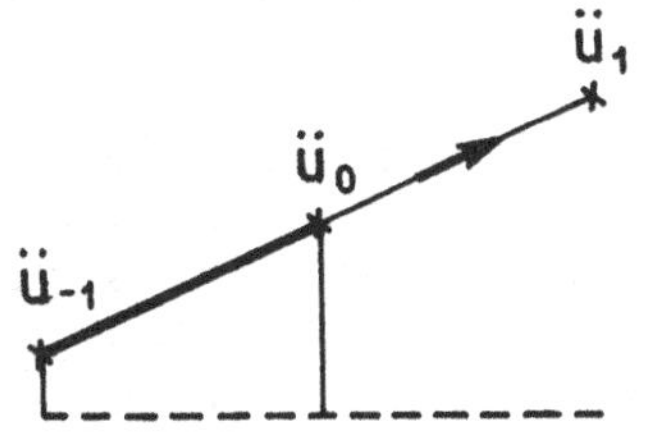

$$\ddot{u}_1 = (2\underset{\sim}{u}_1 - 5\underset{\sim}{u}_0 + 4\underset{\sim}{u}_{-1} - \underset{\sim}{u}_{-2}) / \Delta t^2$$

extrapolating velocity similarly

$$\dot{\underset{\sim}{u}}_1 = \frac{3}{2}\dot{\underset{\sim}{u}}_{\frac{1}{2}} - \frac{1}{2}\dot{\underset{\sim}{u}}_{-\frac{1}{2}}$$

$$\dot{\underset{\sim}{u}}_1 = (3\underset{\sim}{u}_1 - 4\underset{\sim}{u}_0 + \underset{\sim}{u}_{-1}) / 2\Delta t$$

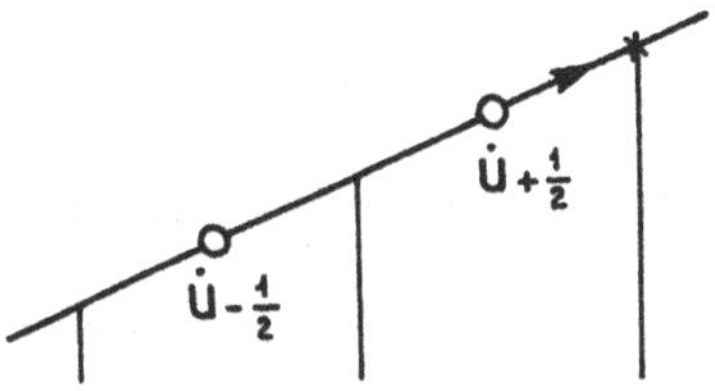

Writing governing equation at time

$$\ddot{\underset{\sim}{u}}_1 = -M^{-1}(\underset{\sim}{C}\,\dot{\underset{\sim}{u}}_1 + \underset{\sim}{R}_1(u) + \underset{\sim}{F}_1).$$

From above

$$\underset{\sim}{u}_1 = \frac{5}{2}\underset{\sim}{u}_0 - 2\underset{\sim}{u}_{-1} + \frac{1}{2}\underset{\sim}{u}_{-2} - \frac{\Delta t^2}{2}\underset{\sim}{M}^{-1}\left[\underset{\sim}{R}_1 + \underset{\sim}{F}_1 + \underset{\sim}{C}\,\dot{\underset{\sim}{u}}_1\right]$$

Equations 2) and 4) solved iteratively.

Also writing $\underset{\sim}{R}_1 = \underset{\sim}{R}_0 + \underset{\sim}{K}_{T0}(\underset{\sim}{u}_1 - \underset{\sim}{u}_0)$ and explicit form can be found.

$$\underset{\sim}{u}_1 = -\left[2M - \frac{3}{2}\Delta t\, \underset{\sim}{C} + \Delta t^2 \underset{\sim}{K}_T \right]^{-1} \left[\Delta t^2 \left(\underset{\sim}{R}_0 + \underset{\sim}{F}_1 \right) + \right.$$

$$\left. + \left(5\, \underset{\sim}{M} - 2\Delta t\, \underset{\sim}{C} \right) \underset{\sim}{u}_0 - \left(4\, \underset{\sim}{M} - \Delta t\, \underset{\sim}{C}/2 \right) \underset{\sim}{u}_{-1} + \underset{\sim}{M}\, \underset{\sim}{u}_{-2} \right] .$$

Explicit form involves re-inversion of a certainly non-diagonal matrix which varies with each step and can be very expensive. Iterative approach can use diagonality of $\underset{\sim}{M}$ matrix but full advantage of assembly may be lost as calculation of $\underset{\sim}{R}_1$ involves all displacements.

However for non-linear cases the iteration is still economic. Method is unconditionally stable and quite large time intervals can be used. Modes with high periods rapidly damped out artificial damping). This is no disadvantage in many practical problems where this in fact happens physically.

A possibility for speeding iteration is to use $\underset{\sim}{u}_1$ predicted by central difference expression as the first step of iteration.

Special assumptions may have to be introduced at start to get

$$\underset{\sim}{u}_{-1} , \underset{\sim}{u}_{-2} .$$

Other backward difference schemes can be used using values of $\underset{\sim}{R}_0$ and $\underset{\sim}{F}_0$ only but extrapolating $\underset{\sim}{u}_1$ from $\dot{u}$, $\ddot{u}$, u values known at 0 - 1 - 2 etc.

M. Hartzman and J. R. Hutchinson [30] use this in non-linear dynamic problems.

Concluding Remarks

In this set of brief notes some indication of the variety of processes to which the finite element method has been applied is presented. Following lectures by other contributors elaborate some aspects in more detail but clearly possibilities of application are limitless. We leave the further development to reader's own efforts.

REFERENCES

[A] O. C. Zienkiewicz: "The Finite Element Method in Engineering Science", McGraw-Hill, London, 1971.

[1] A. Hrenikoff: "Solution of Problems in Elasticity by the Framework Method", J. Appl. Mech. 8, A169--175, 1941.

[2] D. McHenry: "A Lattice Analogy for the Solution of Plane Stress Problems", J. Inst. Civ. Eng. 21, 59-82, 1943.

[3] J.H. Argyris: "Energy Theorems and Structural Analysis", Butterworth London, 1960. (Reprinted from Aircraft Eng. 1954-55).

[4] M.J. Turner, R. W. Clough, H. C. Martin and L. J. Topp: "Stiffness and Deflection Analysis of Complex Structures", J. Aero. Sc. 23, 805-823, 1956.

[5] R. W. Clough: "The Finite Element Method in Structural Mechanics", Stress Analysis Ch. 7, ed. by O. C. Zienkiewicz and G. S. Holister, J. Wiley & Son, 1965.

[6] B. Fraeijs de Veubeke: "Displacement and Equilibrium Models in the Finite Element Method", Stress Anal. Ch. 9, ed. by O. C. Zienkiewicz and G. S. Holister, J. Wiley & Son, 1965.

[7] R. H. Gallagher: "A Correlation Study of Methods of Matrix Structural Analysis", AGARDograph 69, Pergamon Press, 1962.

[8] O. C. Zienkiewicz and Y. K. Cheung: "The Finite Element Method for Analysis of Elastic Isotropic and Orthotropic Slabs", Proc. Inst. Civ. Eng. 28, 471-488, 1964.

[9] O. C. Zienkiewicz and Y. K. Cheung: "Finite Elements in the Solution of Field Problems", Engineer 200, 507-510, Sept. 1965.

[10] E. L. Wilson and R. E. Nickell: "Application of Finite Element Method to Heat Conduction Analysis" Nuclear Eng. and Design 3, 1-11, 1966.

[11] R. Courant: "Variational Methods for the Solution of Problems of Equilibrium and Vibration" Bull. Am. Math. Soc. 49, 1-23, 1943.

[12] W. Prager and J. L. Synge: "Approximation in Elasticity Based on the Concept of Function Space", Quart. Appl. Math. 5, 241-269, 1967.

[13] B. Fraeijs de Veubeke and O. C. Zienkiewicz: "Strain Energy Bounds in Finite Element Analysis by Slab Analogy", J. Strain Analysis 2, 265-271, 1967.

[14] T. H. H. Pian: "Derivation of Element Stiffness Matrices by Assumed Stress Distribution", J. AIAA 2, 1232-1336, 1964.

[15] T. H. H. Pian and P. Tong: "Basis of Finite Element Methods for Solid Continua", Int. J. Num. Meth. in Eng. 1, 3-28, 1969.

[16] O. C. Zienkiewicz and C. J. Parekh: "Transient Field Problems - Two- and Three-dimensional Analysis by Iso-parametric Finite Elements", Int. J. Num. Meth. in Engr. 2, 61-71, 1970.

[17] B.A. Szabo and G.C. Lee: "Derivation of Stiffness Matrices for Problems on Plane Elasticity by Galerkin Method", Int. J. Num. Meth. Eng. 1, 301-310, 1969.

[18] S.G. Hutton and D.L. Anderson: "Finite Element Method - a Galerkin Approach", Proc. Am. Soc. Civ. Eng. 97 EM5, 1503-1519, 1971.

[19] J.T. Oden: "A General Theory of Finite Elements: I-Topological Considerations, II Application", Int. J. Num. Meth. Eng. 1, 205-221; 247-260, 1969.

[20] W.P. Doherty, E.L. Wilson and R.L. Taylor: "Stress Analysis of Axisymmetric Solids Utilizing Higher Order Quadrilateral Elements", Struct. Eng. Lab. Rep., Univ. of California, Berkeley, 1969.

[21] O.C. Zienkiewicz, R.L. Taylor and J.M. Too: "Reduced Integration Technique in General Analysis of Plates and Shells", Int. J. Num. Meth. Eng. 3, 275-290, 1971.

[22] S.F. Pawsey: "Improved Numerical Integration of Thick Shell Finite Elements", Int. J. Num. M. Eng. 3, 575-586, 1971.

[23] D.S. Griffin and R.B. Kellog: "A Numerical Solution of Axially Symmetrical and Plane Elasticity Problems", Int. J. Solids Structures 3, 781-794, 1967.

[24] R.S. Sandhu and E.L. Wilson: "Finite Element Analysis of Seepage in Elastic Media", J. Eng. Mech. Div., Proc. ASCE 95, 641-651, 1969.

[25] L.R. Herrmann: "Elasticity Equations for Incompressible or Nearly Incompressible Materials by a Variational Theorem", AIAA J. 3, 1896-1900, 1965.

[26] G. C. Nayak and O. C. Zienkiewicz: "Elasto-plastic Stress Analysis - A Generalization for Various Constitutive Relations Including Strain Softening" Int. J. Num. Meth. Eng. 5, 113-135, 1972.

[27] C. A. Anderson and O. C. Zienkiewicz: "Spontaneous Ignition. Finite Element Solutions for Steady and Transient Conditions", L. A. 5037, Los Alamos Scient. Laborat. University of California 1973.

[28] O. C. Zienkiewicz, S. Valliappan and I. P. King: "Elasto-plastic Solutions of Engineering Problems. Initial Stress, Finite Element Approach", Int. J. Num. Meth. Eng. 1, 75-100, 1969.

[29] L. Morino, J. W. Leech and E. A. Witmer: "An improved Numerical Calculation Technique for Large Elastic-plastic Transient Deformations of Thin Shells: Part 1 Background and Theoretical Formulation; Part 2 Evaluation and Applications", J. Appl. Mech. 38, 423-428; 429-436, 1971.

[30] M. Hartzmann and J. R. Hutchinson, Lawrence Radiation Laboratory Report UCRL-72728, 1970.

Contents

Printed in the United States
By Bookmasters